Flaviane Ramos dos Santos

Centrality, commercial hubs and consumption

Flaviane Ramos dos Santos

Centrality, commercial hubs and consumption

A study of car dealerships in Presidente Prudente (SP)

ScienciaScripts

Imprint
Any brand names and product names mentioned in this book are subject to trademark, brand or patent protection and are trademarks or registered trademarks of their respective holders. The use of brand names, product names, common names, trade names, product descriptions etc. even without a particular marking in this work is in no way to be construed to mean that such names may be regarded as unrestricted in respect of trademark and brand protection legislation and could thus be used by anyone.

Cover image: www.ingimage.com

This book is a translation from the original published under ISBN 978-3-330-77265-6.

Publisher:
Sciencia Scripts
is a trademark of
Dodo Books Indian Ocean Ltd. and OmniScriptum S.R.L publishing group

120 High Road, East Finchley, London, N2 9ED, United Kingdom
Str. Armeneasca 28/1, office 1, Chisinau MD-2012, Republic of Moldova, Europe
Managing Directors: Ieva Konstantinova, Victoria Ursu
info@omniscriptum.com

Printed at: see last page
ISBN: 978-620-8-64313-3

SUMMARY

ACKNOWLEDGMENTS

As solitary as academic work may seem, it is actually the result of a collective construction, where those people who are part of our lives and who are, at various times, supporting us are very important, mainly because it is precisely they who help us to renew our breath to continue our studies and academic research. For this reason, I would like to thank all those who, directly or indirectly, contributed to the completion of this work and, consequently, to my academic training.

I would first like to thank the São Paulo State Research Foundation (FAPESP) for funding both the research carried out in Presidente Prudente (Process 2013/08710-7) and the research internship in Porto, Portugal (Process 2013/23185-6), which gave me new perspectives of analysis and experience as a researcher.

I would like to thank my advisor, Maria Encarnaçao Beltrao Sposito (Carminha), for her example as a researcher, teacher and person, for her guidance and for all the attention she gave during the construction of this work.

I would like to thank my family: my parents - Selma and Vlademir - my brothers and sister-in-law - Fernanda, Flâvio and Thais - and my nephews - Vinicius, Mariana, Guilherme, Lucas and Beatriz, for the moments of relaxation and renewed energy.

Finally, I would also like to thank my friends, especially Hiago and Carolina, and Claudinei, for their support during the course of this work.

Thank you all so much!

INTRODUCTION

Population and territorial growth, as well as the city's economic development, do not happen uniformly, either in time or in space. With regard to the distribution of uses of urban space, we can see areas that are less endowed with means of collective consumption and economically devalued and, on the other hand, better-equipped sectors that are constantly renewing and restructuring their uses based on the interests of the agents who shape urban space (CORRÊA, 1989), which can even be contradictory to each other. When these dynamics are observed over time, the redefinition of these uses according to different areas of the city is remarkable, revealing the process of structuring the city's space.

Commercial and service activities, and the consumption associated with them, play a fundamental role in this process, since the areas where these activities are concentrated, whether specialized or not, form important nodes that guide the distribution logic of other land uses, especially residential. There has historically been a tendency for new commercial and service areas to appear in association with the population and territorial growth of cities, but what is new are the logics that guide this trend in contemporary times.

In recent years, there has been a strong growth in demand for vehicles, stimulated by the expansion of production, relative growth in workers' incomes and the recent ease of access to credit, encouraged by recent governments. In this context, the automobile has become an essential element of circulation in urban spaces and has contributed to substantial changes in the structure of cities, especially with regard to the reproduction and expansion of these spaces, since the automobile has made it possible to move around the city more easily, thereby increasing the radius of action of consumers.

Another important issue that we cannot fail to mention is the social '*status*' that is attributed to owning a car in contemporary times. This is reinforced by the lack of quality and efficiency in public transport, as well as media *marketing*, which plays a fundamental role in spreading the 'need' for a car in people's lives.

In this context, the relative growth in workers' incomes and the ease of access to credit, combined with the Brazilian federal government's initiative to reduce the Tax on Industrialized Products (IPI) for cars, has a very important role to play in stimulating consumption while bringing about a relative reduction in prices. Remembering that this measure to reduce the IPI rate is mainly justified by the importance of the automotive sector for the economy as a whole, since it represents around 20% of the country's industrial production, according to the National Association of Automotive Vehicle Manufacturers (ANFAVEA, 2014).

Bearing these premises in mind, the aim of this text is to analyze the process of concentration of car dealerships on three important axes of circulation in the south-southwest sector of the city of Presidente Prudente (SP), as well as to identify the locational choices of these establishments and their relationship with the action of urban space-shaping agents.

The city of Presidente Prudente (Map 1) is located at latitude 22° 7'S, longitude 51° 22'W and has an altitude of around 480m above sea level. It is also the largest city in terms of population in the far west of the state of São Paulo, with 207,625 thousand inhabitants, according to the Demographic Census of the Brazilian Institute of Geography and Statistics (IBGE, 2010).

Map 1

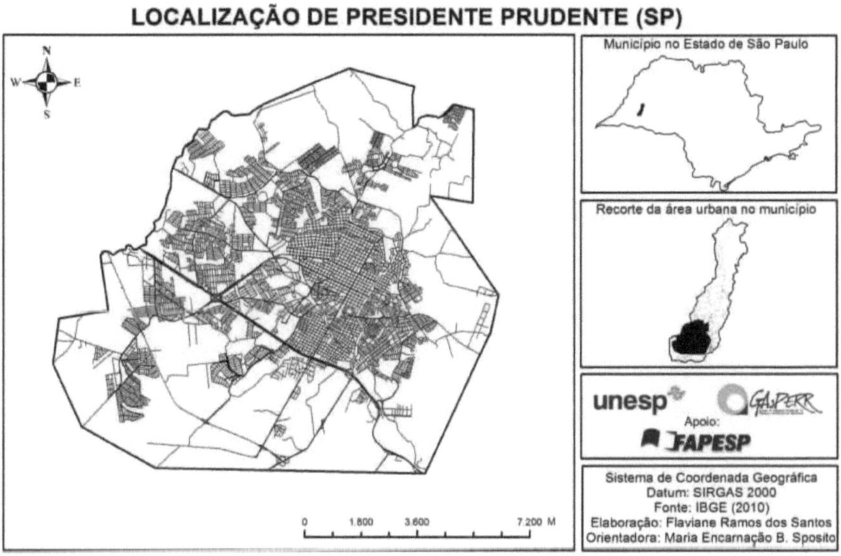

Source: IBGE (2010).

To this end, the research was based on reading and reviewing texts on the Brazilian automotive industry and the production of urban space, with an emphasis on themes that dealt with concentration and functional and/or socio-economic specialization, emphasizing the emergence and strengthening of new commercial and service areas, reconstituting urban centrality on multiple scales.

Secondly, a search was made for information in old telephone directories, in order to look back and record the location of establishments specializing in car sales in Presidente Prudente. After *downloading* the *shape of* the urban grid of Presidente Prudente from the IBGE (Brazilian Institute of Geography and Statistics) website, using *Google Maps* to locate the addresses and *ArcGis* to record the location of the establishments, three maps were made for the years 1988, 1998 and 2008.

Next, an interview script was drawn up to be carried out with the managers of the concessionaires. The interviews were then carried out in the field and, with the help of a GPS (*global* positioning system), the geographic coordinates of the concessionaires were collected so that it was possible to make a map in *ArcGis* of their location in Presidente Prudente for 2014.

In order to provide an overview of the automotive industry in Brazil, research was carried out and files *downloaded* from the *websites* ANFAVEA, the National Federation of Motor Vehicle Distribution (FENABRAVE) and the National Traffic Department (DENATRAN), as well as searching for news about the sector.

In order to interpret and systematize the data, both from the aforementioned institutions and from the interviews and fieldwork, tables, graphs and maps were produced using *software* such as *Excel* and *ArcGis*. Finally, once the data had been obtained, systematized and represented in cartographic form, the paper was finally written.

This paper refers to the research funded by FAPESP and associated with the thematic project: "Economic logics and contemporary spatial practices: medium-sized cities and consumption", which was developed between August 2013 and January 2015, and is structured in four chapters, in addition to this introduction.

In the chapter entitled **Panorama of the automotive industry in Brazil**, as the title suggests, a brief analysis of the sector is presented, taking into account total production data, the sector's share of industrial GDP, exports, investment and jobs generated, highlighting the importance of the industry for the country's economy, considering it as a driving industry, as Kon (1994) points out. In addition, data is presented on the evolution of the car fleet in the country and in Presidente Prudente.

Next, in **The production of urban space - Discussing** concepts, a discussion is presented, based on various authors, of the concepts of urban network, average city, urban and city restructuring, center and subcenter, the concept of centrality and its variations, such as changing centrality, polynucleated centrality, multicentrality and planned centralities, as well as the concepts of segregation and fragmentation, associating them, at various times, with the locational choices of car dealerships in urban space.

In the chapter entitled **Car dealerships in Presidente Prudente**, we first describe the characterization of the city chosen as a sample and, secondly, based on Lima (2009), the form of organization and, above all, what we are considering in this work to be a dealership and, finally, a brief description of each of the thirteen dealerships in the city, as well as the groups that manage them.

Last but not least is the chapter on the **structuring of urban space and the spatial logic of car dealerships,** divided into three parts. In the first part, based on information obtained from old telephone directories provided by the EPIL publishing house, the history of the location of car sales establishments in the city is presented, as well as the changes in their locational choices over the 20 years chosen for analysis. In the second part, maps, graphs and tables are presented, representing information described by the interviewees, as well as some excerpts from these interviews. In the third part, the strategies adopted by car dealerships in Presidente Prudente with regard to their location choices are listed.

Finally, we have the **Final Considerations** on the research carried out, as well as its importance for understanding the process of urban restructuring and the city in the contemporary period.

The research on which this book is based is associated with the thematic project: "Economic logics and contemporary spatial practices: medium-sized cities and consumption" (Process 2011/20155-3), which contributed to its development, as well as benefiting from the work being carried out by the whole team.

1. OVERVIEW OF THE AUTOMOTIVE INDUSTRY IN BRAZIL

The articulation between the different geographic scales is very important in order to understand a particular phenomenon, dynamic or process in space, thus relating the various levels, from the national scale, through the urban network to the space of cities, remembering that both, to some extent, are also articulated with the international scale. On this issue, Sposito (2011, p.130) points out that "nothing can be explained on just one scale, even if we are referring to the international scale".

Therefore, in order to understand the locational choices made by dealerships in Presidente Prudente, it is essential to analyze the national context, as well as the context of the urban network in which Presidente Prudente is located, in view of its role as a polarizer over a large group of small towns, taking into account that the purchase of vehicles is a type of occasional consumption, which justifies traveling longer distances to make it happen.

Following this premise, in this chapter we aim to carry out a brief analysis of the automotive industry in Brazil, given that it is a sector of national and, in turn, global prominence, as well as being the target of various public policies, among other governmental actions, which ensure and boost its growth, and is also influential in various other branches of production and services, which makes it a driving industry.

According to Kon (1994, p.171), a driving industry "has the property of, by increasing the volume of its production, causing an increase in the volume of production of various so-called 'driven' industries." In other words, the automotive sector involves a broad and significant participation of sub-sectors with *backward effects*, which "consist of the interrelationships that occur, respectively, with industries supplying inputs and other services necessary for the production process", and *forward effects* in the production chain, which "are related to the distribution and repair industries of the finished product". (KON, 1994, p.171-72). Thus, a traditional example of a driving industry is the automotive industry, as the author puts it, because:

> [...] its 'backward' multiplier effects attract the establishment of auto parts companies and other specific inputs, and its 'forward' effects attract the establishment of car dealerships, consortia, gas

stations, maintenance services, car finance companies, etc. (KON, 1994, p. 172).

The automotive industry is made up of two major complexes responsible for producing cars: the automakers and the auto parts suppliers. Graph 1 shows that the sector has great national importance and, although there have been some fluctuations, the sector has shown an upward trend in its share of the industrial Gross Domestic Product (GDP), reaching 18.70% of the total wealth produced in the country in 2012.

Graph 1

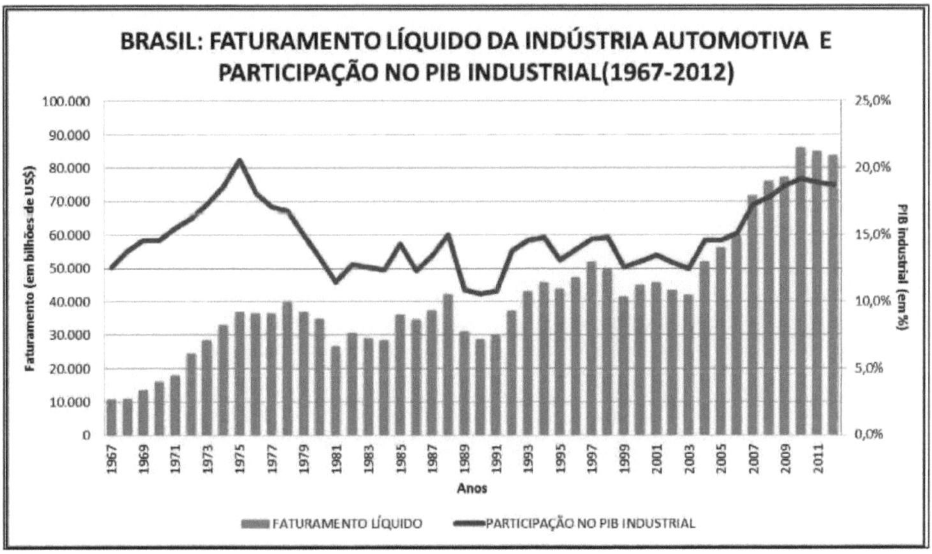

Source: ANFAVEA (2014) - Org. Flaviane Ramos dos Santos.

However, the forecasts of the National Association of Motor Vehicle Manufacturers (ANFAVEA) and the National Motor Vehicle Distribution Federation (FENABRAVE) for 2014 are for a probable drop in this percentage, according to FENABRAVE's 2013 Yearbook, p.72:

Publishing the 2013 Yearbook, we can see that the reality of the sector is no longer the same as in past years, when we recorded successive increases, often in double digits. The automobile segment has shrunk, for different reasons [...] For 2014, caution affects all segments, due to the greater complexity that the year brings, as a result of the lower volume of sales days, with Carnival in March, the World Cup between June and July, and the Elections in the second half, as well as the volatility of the economy, which predicts GDP growth below 2%, rising interest rates, the threat of inflation, among other reasons.

In addition, the automotive industry has been going through a period of falling exports, above all because of the crisis in Argentina, which has always been one of the main destinations for domestic vehicles. According to Costa (2014), exports fell by 13.3% in the first quarter of 2014 compared to the same period last year and by 45.2% compared to the last quarter of 2013. Graph 2 shows an upward trend in car exports in Brazil, especially from the 2000s onwards. After reaching its peak in 2005, there was a period of instability with a significant drop in 2009, a resumption of growth in 2011, a further drop in 2012 and a small increase in exports in 2013, with these fluctuations largely accompanied by fluctuations in total car production in the country.

Graph 2

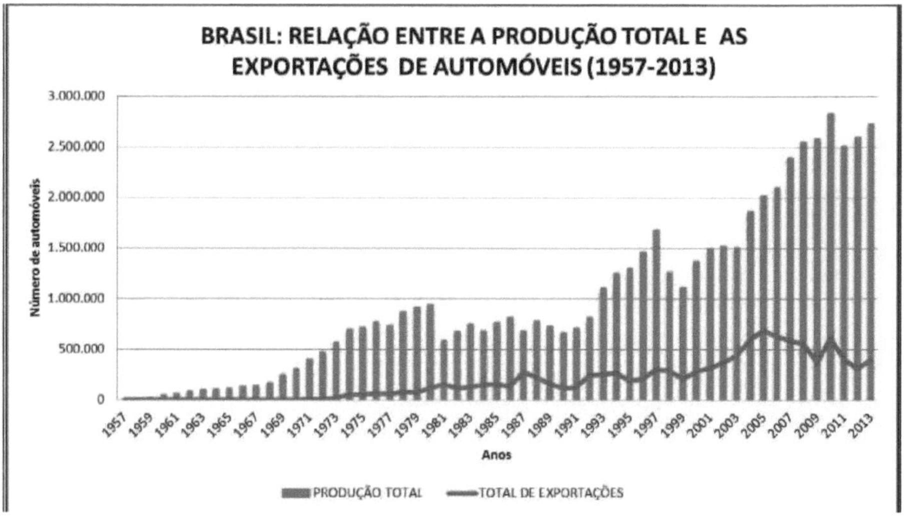

Source: ANFAVEA (2014) - Org. Flaviane Ramos dos Santos.

The importance of the sector for the country's economy is also notable due to the total number of direct jobs generated, as shown in Graph 3, where we can see a period of continuous growth after 2003, with slight decreases in 2005 and 2009. In 2013, considering only the generation of direct jobs[1] in vehicle production, the automotive

1 According to Luna (2013), the advance of new labor management practices resulting from productive restructuring, such as outsourcing and other types of flexibilization, means that the number of workers involved

industry reached a total of almost 132,000 workers, something that hadn't happened since 1979 and 1985.

Graph 3 - Brazil: Evolution of employment in the automotive industry (1957-2013)

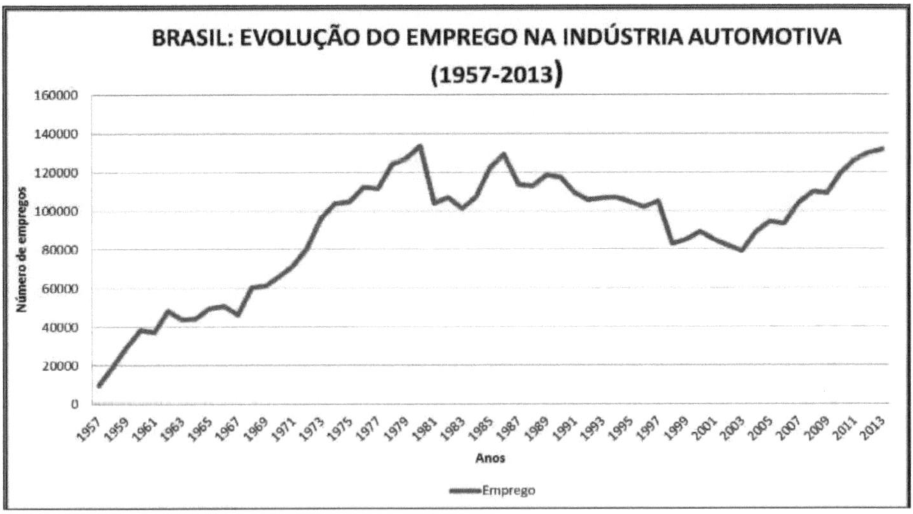

Source: ANFAVEA (2014) - Org. Flaviane Ramos dos Santos.

However, the recent downturn in the automotive industry has led automakers to adopt measures to adapt production to market demand, such as temporary contract suspensions *(layoffs)*, Voluntary Dismissal Programs (PDV), collective vacations and reductions in working hours. According to ANFAVEA, in November 2014, the level of employment in the industry fell by 6.9%, i.e. 9,406 thousand fewer workers than in the same month in 2013.

On this issue, we can mention the *Volkswagen* workers in Sao Bernardo do Campo[2] who went on strike on January 6, 2015 in protest at the 800 layoffs announced through letters sent to employees during the period they were on collective vacation. According

certainly exceeds these figures, since the automotive industry (understood as a motor industry) requires a considerable production chain (parts and other service providers).

2 *Volkswagen* has four plants in Brazil. In Sao Bernardo do Campo (SP), which opened in 1959, it has a total area of 1,000,000 m2 and 13,000 employees. The second unit was opened in 1976 in the city of Taubaté (SP) and has a total area of 290,000 m2, with around 5,000 employees. Inaugurated in 1996, its third unit is located in Sao Carlos (SP) and has a total area of 84,000 m2, with around 800 employees. Finally, its fourth unit is located in Sao José dos Pinhais (PR) and has a total area of 300,000 m2, with around 3,300 employees, having been inaugurated in 1999, according to the carmaker's official *website*.

to a report published in the newspaper O Estadao, "*Volkswagen* claims to have a surplus of 2,000 workers, out of a total of 13,000 at the plant, and also said that this is the first stage of adapting the workforce.[3] "

After ten days of strike action, which meant 14,000 fewer cars being produced, *Volkswagen* agreed to reinstate the 800 dismissed employees, thus ending the strike, according to a report published by Folha de Sao Paulo .[4]

With regard to investments made in the sector (Figure 1), we can see a decline between 1990 and 1993, followed by a sharp rise after 1994, which continued practically until 1998, culminating in a significant drop in subsequent years, with recovery only from 2005 onwards.

Figure 1 - Brazil: Investments in the automotive sector (1980-2012)

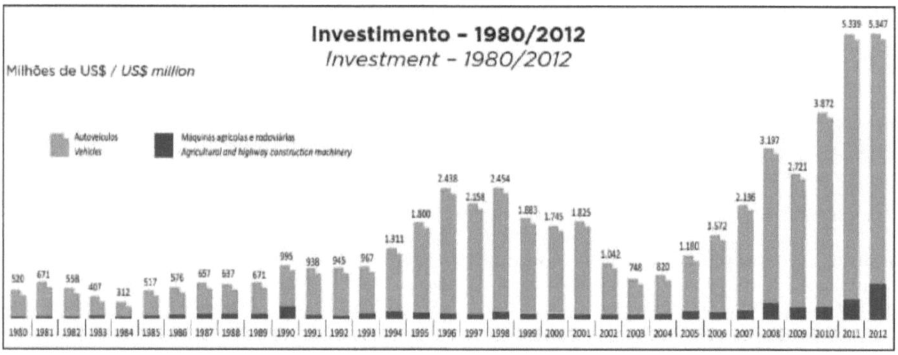

Source: ANFAVEA, 2013.

It is worth considering the leap in investment after 2005, sustained by the industrial policies of the last few governments which, through wage increases, social welfare programs, a reduction in the tax burden (for example, the reduction in the IPI tax rate, which will be discussed later) and access to credit, allowed for a substantial rise in car consumption.

The National Traffic Department (DENATRAN) has reported that the private car fleet

---

3 Available at: <http://economia.estadao.com.br/noticias/geral,empregados-da-vw-decretam-greve-e-governo-descarta-ajuda para-o-sector,1616107> Accessed January 2015.

4 Available at:< http://www1.folha.uol.com.br/mercado/2015/01/1575970-volkswagen-readmite-800-funcionarios-e-greve-de-10-dias-acaba.shtml> Accessed on: January 2015.

has grown by 85% in Brazil in the last 10 years, while in the same period the population has increased by 6.5%. In terms of numbers, the car fleet amounts to approximately 45.4 million, according to DENATRAN data, and the absolute majority of these cars are concentrated in metropolitan areas and present a huge obstacle to urban mobility and the quality of life of city dwellers. On the other hand, the figures show that the big cities are maintaining their car fleets, while the growth rates of the car fleet in small and medium-sized cities are increasing at galloping proportions. In other words:

Car sales in Brazilian cities with more than 10 million inhabitants grew by just 6% between 2007 and 2013, a period in which sales in the country grew by 53%, according to a study by ANFAVEA. [...] Sales grew below average in all 638 municipalities with more than 500,000 inhabitants. In other words: the big cities are exhausted, but we won't reach chaos, because they have the same profile as European capitals, American and Japanese cities, that is: growth is vegetative, there is a replacement market. So where did all the cars licensed during this period go? To small towns, where **the smaller the town, the greater the growth**. Note that the 1,252 Brazilian municipalities with less than 5,000 inhabitants have seen an astonishing increase in car sales: 142% in the last six years (remember that the total market grew by only 53% in the period). Cities with up to 10,000 inhabitants grew by 124% and sales also increased in cities with up to 500,000 inhabitants. (Article published in ABRAC - Brazilian Association of Chevrolet Dealers - 2014) - Emphasis added.

Analyzing the evolution of the car fleet in the city of Presidente Prudente, the city chosen as the territorial focus of this study, it was found that the fleet has doubled in the last twelve years, from 46,245 cars in 2002 to 86,152 in 2014, according to data from DENATRAN. Furthermore, when this data is compared with population growth, the situation is worrying, as population growth in Presidente Prudente between 2000 and 2010 was 8.8% (IBGE, 2010), while the increase in the car fleet was 38.3% between 2002 and 2012[5] (DENATRAN, 2014). In Graph 4, we can see that the evolution of the car fleet in Presidente Prudente is largely in line with the evolution of the fleet in Brazil, with both showing an upward trend.

Graph 4

5 Comparison over the same period was not possible, as data on the car fleet in Presidente Prudente in 2000 is not available on the DENATRAN *website*.

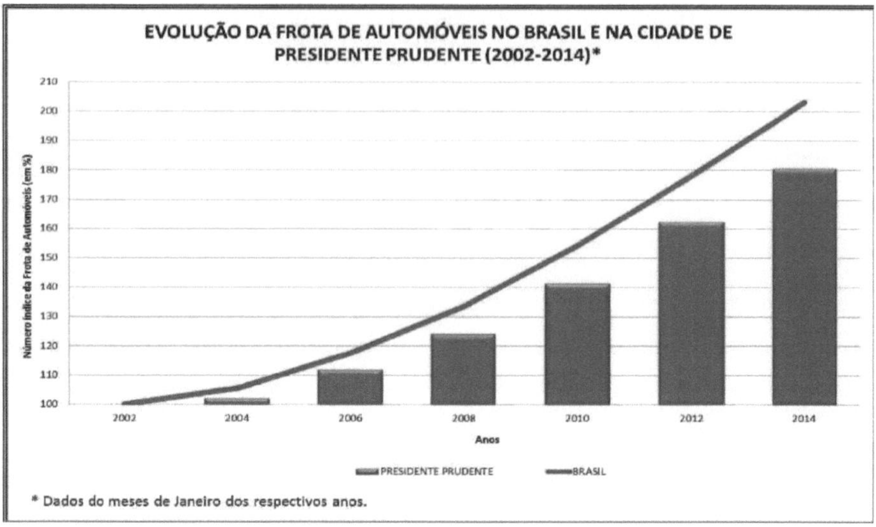

*Values used are index numbers .[6]

Source: DENATRAN - Org. Flaviane Ramos dos Santos.

Thus, in recent years, there has been a strong growth in demand for cars, stimulated by the expansion of production, relative growth in workers' incomes and the recent ease of access to credit, encouraged by recent governments.

Thus, given the importance of the automotive industry for national economies, as it is a driving industry, as mentioned above, most governments develop specific policies for the sector, such as voluntary restriction agreements, structural barriers, quantitative quotas or rules of origin, as well as taxing imports (LOPES, 2006). According to the same author, despite globalization and pressure from multilateral trade bodies, in practice what has happened is that specific policies have been maintained for the sector, protecting it from greater exposure to competition.

Brazil is no different. The history of the federal government's support for the automotive industry is historical, as this sector has benefited from fiscal and financial incentive policies since the 1950s. In this context, it is undeniable that these stimulus

6 The use of index numbers makes it easier to calculate cumulative percentage changes between certain periods of time. They have a base equal to 100 in a given period and from then on you simply apply the percentage change obtained in the following month to the previous month's index number.

policies have led to the growth of the sector in recent decades.

As we already know, the choice of road transportation over other modes (which would be more efficient for a country of continental dimensions like Brazil) dates back to the government of Juscelino Kubitschek in the 1950s. And this happened because of the demands of the automotive industries themselves, given that in order for them to be installed in the country it was necessary: firstly, to create the need for a car in the daily lives of the population (in this respect, advertising and *marketing* played a fundamental role) and, secondly, to create the necessary conditions for these cars to move around the country, in other words, the highways.

Furthermore, when we talk about the automotive sector, we have to consider the impact of the entire production chain on the economy and not just the automakers. In other words, the sector has ramifications and impacts on different segments of the economy (rubber, plastics, steel and electronics, for example) and is therefore considered by the state to be a strategic sector.

The Brazilian Federal Government's initiative to reduce the Tax on Industrialized Products (IPI) for cars from 2009 onwards is a measure whose main justification is the importance of the sector for the economy as a whole, since it accounts for around 18% of the country's industrial production (ANFAVEA, 2014), in addition to the clear aim of boosting consumption and stimulating the economy.

According to Buffon (2011), *it* should be recognized that one of the important instruments used by the government to deal with the economic crisis of 2008 was the reduction in the tax burden, which meant that Brazil did not see significant unemployment rates, such as those seen in other countries most severely affected by the crisis (BUFFON, 2011). At the same time, we can observe that the privileges of the automotive sector in the country are the result of its political strength (both on the part of businessmen and workers) and its exposure in the media.

Therefore, the automotive sector is very important for the economy and therefore deserves special attention, because as well as representing a significant percentage of GDP, it is an industrial sector with high quality standards, so that it can stand up to international competition. In other words, it is possible to see the importance of the automotive sector in Brazil, reflected in the number of public policies, among other strategies, that enable the growth of the automobile production chain, a sector that accounts for a considerable share of the country's industrial GDP and also generates direct and indirect jobs.

2. THE PRODUCTION OF URBAN SPACE - DISCUSSING CONCEPTS

São Paulo's urban network is, among others, the most important in terms of demographic size and the presence of cities of different sizes. The development of the coffee economy, which influenced the formation of this network, its consolidation with the industrialization process and its diversification in terms of the growing role of consumption are factors to be considered. Corrêa (2006, p. 311) defines the **urban network** as:

> [...] a set of functionally articulated centers, which is a social reflection, the result of complex and mutable processes engendered by various social actors. A variety of types of urban networks emerge from this complexity, varying according to combinations of characteristics such as the size of the centers, their density in the regional space, the functions they perform, the nature, intensity, periodicity and spatial scope of the interactions and the shape of the network.

Camagni (2005) and Catelan (2013) discuss the overlapping of two urban network patterns. The first is the hierarchical pattern, which occurs when small cities are subordinate to medium-sized cities and medium-sized cities to large cities, i.e. there is a vertical articulation between cities. The second pattern is what Camagni (2005) calls networks structured by horizontal relationships between similar or dissimilar urban centers, while Catelan (2013) calls this second pattern heterarchical, the result of inter-scalar spatial interactions in the urban network. According to him,

> The hierarchical pattern continues to exist, but it has become insufficient to explain the contents arising from the increase in spatial interactions under the aegis of globalization. The nodes of the networks are growing not only in number, but also in technical, territorial and economic complexity, with greater synergy between the roles of each city in the urban network, at each geographical scale and in the interaction between them (CATELAN, 2013, p. 15).

In other words, this new pattern (heterarchical), which overlaps and articulates with the previous pattern (hierarchical), causes a **restructuring of relations between cities** which, according to Sposito (2011), can be seen from two perspectives: the intensification of relations within different urban networks and the possibility and realization of spatial interactions between cities belonging to different urban networks.

Interest in the study of medium-sized cities, and especially medium-sized cities, emerged in the mid-1970s, as their relative importance in the Brazilian urban network began to grow. The diseconomies of agglomeration associated with metropolitan areas, the disjunction between management and production activities, as well as the lower costs of producing urban space, among other factors, have played an important role in redefining the roles of these cities in the interurban division of labor.

The concept of a **medium-sized city** is very complex and can be defined on the basis of two criteria. The first is population, where for a city to be considered medium-sized it must have a population range between 100 and 500 thousand inhabitants. However, identifying a city based on the size of its population is limiting and, for this reason, some authors prefer to use the term **medium-sized** city to refer to cities with this population range that do not play the role of a medium-sized city (SPOSITO, 2001b).

The second and more common criterion is the functional one, which takes into account the relationships that the roles played by these cities provide. In other words, it is of the utmost importance to consider the intermediary roles played by the city in the context of the urban network as a link between the metropolis and the small town, or even, according to Sposito (2011, p.51) to consider these as being "tributaries of the metropolis and polarizers of the small towns that are in their area of direct influence".

Presidente Prudente, considered a medium-sized city from the point of view of population and, above all, function, follows the general logic of city production under the aegis of capitalism, which results in a very unequal configuration. According to Pereira (2002, p. 1), this logic is based on:

> [...] discontinuous territorial growth with the occurrence of horizontal and vertical expansion; the occurrence of numerous urban voids; the implantation of housing estates and condominiums closed in the periphery; the multiplication of central areas, as well as the constitution of sub-centers and specialized roads; the unequal distribution/concentration of commercial equipment, services, infrastructure and urban equipment.

At the same time, we must consider that the city is historically produced by different agents who often act in a contradictory way, because they have different objectives and power to act, in terms of access to the best locations and their implementation over time (CORRÊA, 1989). According to the same author, the social agents who play a fundamental role in the process of producing urban space are: the owners of the means of production, landowners, real estate developers, the state and excluded social groups.

So, taking Corrêa's book "O espaço urbano" (1989) as a reference point, let's start by talking about the **owners of the means of production**, who are major consumers of space due to the size of their activities. For this reason, they need large, cheap plots of land that meet the locational requirements of their companies' activities, for example, close to places with wide accessibility, railway lines, airports or port areas.

In the case of large cities where industrial activity is significant, the action of the owners of the means of production interferes in the production of urban space in a significant way, because they create large factory areas in sectors that are different from the prime residential areas (where the population with the highest purchasing power lives), but close to the popular residential areas (where the workers live). In this way, "their actions shape the city, producing their own space and decisively interfering in the location of other land uses. " (CORRÊA, 1989, p. 15).

Landowners are focused on obtaining the highest possible income from their property and are more interested in the exchange value of the land than its use value (SAVÉRIO SPOSITO, 2008).

In order to install urban infrastructure that contributes to the appreciation of their properties, owners of well-located land, valued for physical amenities, such as flat terrain, a river or a panoramic view, for example, put pressure on the municipal authorities. When these lands with amenities and/or externalities are located on the outskirts of cities, they are destined for people with higher purchasing power, where selective neighborhoods and/or high and medium standard horizontal condominiums are created (CORRÊA, 1989). On the other hand, the owners of poorly located land (on the outskirts with no amenities or externalities) are left with the strategy of creating popular housing estates, where the minimum infrastructure is required.

In this way, unlike the owners of well-located land, they have no interest in becoming real estate developers, given the low purchasing power of the population who want to live on these poorly located plots. In this way, popular subdivisions are created on the outskirts, where "[...] floods, mosquitoes, crime and hours lost in precarious public transportation are just some of the aspects of the daily lives of the residents of this place. " (CORRÊA, 1989, p.19)

Real estate developers, in turn, refer to the group of agents who carry out the following operations: a) development; b) financing; c) technical study; d) construction or physical production of the property; e) commercialization or transformation of commodity capital into money capital, the first of which refers to the investment stages, while the last operation is when real estate developers make their profit (CORRÊA 1989).

Faced with the capitalists' lack of interest in producing low-income housing, the big banks and the state itself also act as real estate developers. This occurs for two main reasons: firstly, because of "the low wage levels of the working classes, compared to the cost of capitalistically produced housing" (CORRÊA, 1989, p. 21).[7] The second is due to the conflicting interests of the landowner, the real estate developer and the building materials industry, which has the objective of producing housing with innovations, thus obtaining an ever higher sale price. However, according to the author, it is possible to make the production of housing for low-income groups profitable if they are over-occupied; if they are built with poor quality materials; and when there is a huge shortage of housing, causing prices to rise.

But, as I said earlier, capital has no interest in producing housing for the working classes, since its priority is to produce homes to satisfy solvable demand and, at a later stage, if it sees a profitable advantage, to get help from the state to make it feasible to produce homes for the less well-off. Therefore, "the action of real estate developers in the urban space occurs unevenly, creating and reinforcing residential segregation, so

7 In the current period, there is interest among economic agents in the production of housing for the working classes, based on financing and new ways for the private sector to participate in the production of social housing, through the federal government's "Minha casa Minha Vida" program. In this context, we can highlight companies in the real estate sector such as MRV Engenharia.

characteristic of the capitalist city. " (CORRÊA 1989, p. 21-22)

Another agent that acts in the organization of urban space is the **state**, which has multiple roles, according to Corrêa (1989, p. 24),

> [...] it can act directly as a major industrialist, consumer of space and specific locations, landowner and real estate developer, while also being an agent for regulating land use and the target of so-called urban social movements.

However, it is through the implementation of public services that the state's action is most significant, such as bringing water, sewage, garbage collection, lighting and asphalt to sectors of the city that previously did not have these basic services. With this, the state is contributing to the valorization of this sector and encouraging the installation of commercial and service enterprises, or depending on the case, industrial enterprises, remembering that the state's spatial action occurs unevenly (CORRÊA, 1989).

Another role of the state in shaping city space is to establish the legal framework for the production and use of space (CORRÊA,

2011). In other words, its function is to draw up laws and regulations on land use, among other forms of zoning. With these attributions, the spatial performance of the state does not occur in a socially neutral way, or according to Corrêa (1989 p. 25-26).

> Its action is marked by the conflicts of interest of the different members of class society, as well as the alliances between them. It tends to favor the interests of the segment or segments of the ruling class that are in power at any given time.

The state therefore creates mechanisms that lead to and ratify residential segregation through two main actions, one direct and the other indirect. The indirect action concerns the differentials in property tax (IPTU) and the spatially differentiated allocation of urban facilities, both of which are discriminating factors because they affect the price of real estate and land. The direct and explicit action of the state refers to the creation of popular urban centers, often in discontinuity with the urban network, which causes the population with lower purchasing power to move to these areas of residence (CORRÊA, 1989).

Considering that the state has three political-administrative levels (federal, state and municipal), it should be remembered that "at each level, action changes, as does the discourse that covers up the dominant interests. It is at municipal level, however, that these interests become more evident and the discourse less effective" (CORRÊA, 1989, p. 26).

Last but not least, we have the action of **excluded social groups** as shapers of urban space. Considering that housing is a selective good and therefore the most significant symptom of exclusion, these groups have the following housing options: a) houses produced by the self-build system; b) housing developments produced by the state; c) allotments far from the center; and d) the favela (CORRÊA, 1989).

The first three housing possibilities presuppose a link to a social agent, because none of them transforms the excluded population into a shaper of urban space, since they are subject to the logic of the landowners on the outskirts (CORRÊA, 1989).

In this way, excluded social groups effectively become agents shaping urban space through the production of favelas, whether on public or private land. Remember that,

> The production of this space is, first and foremost, a form of resistance and, at the same time, a survival strategy. [...] Resistance and survival translate into the appropriation of land that is usually unsuitable for other agents in the production of space, such as steep slopes and flooded areas. This is de facto appropriation. On an immediate level, the favela is a solution to a double problem: housing and access to the workplace (CORRÊA, 1989, p. 30).

Therefore, according to Corrêa (1989, p.31), these excluded social groups fight for the right to the city and the implementation of basic urban facilities by the state, which "occurs through the pressure exerted by residents organized in associations, or from electoral interests".

In his article in the book "The production of urban space", Corrêa (2011) discusses the 'distinct spatial strategies and practices of a single agent' and the 'similar spatial strategies and practices of different agents'. In the first case, he refers to the fact that the same agent can operate in different branches of activity, such as a large industrialist who invests his surplus in the production of real estate or in the creation of allotments.

And in the second case, it refers to spatial practices that can be common to different agents, such as "the sterilization of land, fragmentation and remapping, as well as the creation of discontinuous allotments on the periphery. " (CORRÊA, 2011, p. 45).

With regard to the location choices made by car dealerships in Presidente Prudente, we can highlight the actions of landowners, real estate developers and, above all, the state, which, according to Corrêa (1989), has the role of providing the necessary conditions for the actions of other agents, such as improving traffic routes, the sanitation system, among others. We should also point out that city dwellers also produce urban space through their spatial practices, which in a way influences companies' location decisions.

In the case of Presidente Prudente, the municipal government contributed to the increase in value of the area around the extension of Avenida Manoel Goulart, with the canalization of the Saltinho stream[8] , in 2009, with an investment of R$9.5 million, part of which was subsidized by the federal government through the Growth Acceleration Program (PAC). In addition, the resurfacing of the roads and, above all, the improvement in public lighting, inhibiting the presence of prostitutes and transvestites in the vicinity, also contributed to the development of the avenue, which led to the construction of buildings and the installation of various activities linked to commerce and the provision of services.

In this context, taking into account the importance of social agents in the production of urban space, these being "[...] concrete social agents, with rigidly defined roles, carrying interests, contradictions and spatial practices that are sometimes their own and sometimes common" (CORRÊA, 2011, p.41).), we highlight the multiplication of areas where commercial and service activities are concentrated and the redefinition of the urban structure through the emergence of new centralities in the urban space as a result of the actions of these agents.

8 The Saltinho stream is 2,492 meters long and six meters wide, starting at the Sao Joao Baptista Municipal Cemetery and ending near Avenida Manoel Goulart, according to information from the Presidente Prudente City Hall.

In general, the **center** is understood as a place while **centrality** is understood as a condition, which is identified by the diversity of flows and the fluidity of the territory (PEREIRA, 2002). With regard to the urban center, Santos (1989, p. 181) points out that:

"[...] the most striking characteristics are that it constitutes the main node of the urban road network (in this respect, there can be several centers within the same city) and that it has a strong concentration of services at all levels, especially shops."

For Fernandes (2013, p. 31-34) the centers are understood as:

Symbolic places of identity and cohesion of an enlarged and internationalized territory, as spaces of concentration of establishments of activities of greater specialization and area of influence and as essential points of concentration and divergence in the networks of urban circulation [...] territory endowed with high accessibility [...].

For Johnson (1974), the main characteristic of the center is accessibility, i.e. the part of the city that can be reached most easily. In the case of Presidente Prudente, it is the four main avenues that cut through the city, i.e. Avenidas Brasil, Washington Luiz, Coronel Marcondes and Manoel Goulart, that give the center its character as an area of greater accessibility.

Depending on the size of the cities, their economic characteristics and the profile of the population, they can present different types of spaces according to the predominant activity.

In areas where residential use clearly predominates, there is often nothing more or not much more than neighborhood commerce (supply and daily consumption). In some areas, commerce and services are concentrated, presenting themselves as true intra-urban central locations [...] The areas where the most sophisticated goods are produced or marketed are not totally dispersed in the urban fabric, but, quite the opposite, tend to be restricted to a few areas within the city (SOUZA, 2003, p. 64).

The significant population growth that the city of Presidente Prudente has experienced since the 1970s has been accompanied by an increase in its urban space and, consequently, has caused an expansion in the relationship between the center and the periphery, since it is no longer possible to satisfy all of society's needs in the center due to the increase in distances and intensity of flows.

Over the years, this has led to the emergence of areas in the city with a wide range of shops that can serve the population in their immediate vicinity, in other words, the emergence of **sub-centers** of shops and services, which aim to prevent "residents of different neighborhoods from necessarily having to travel to the CBD[9] [main center] whenever they need to buy a more sophisticated good than bread, milk or newspapers" (SOUZA, 2003, p.65).

With the multiplication of central areas, there is a reordering of space, which puts an end to the single centrality, making multicentrality possible, which according to the functional diversity that is established, also generates polycentrality (PEREIRA, 2002).

Sposito (2001a) differentiates between multiple centrality and polynucleated centrality. According to Sposito, **multiple centrality** does not occur through the territorial dispersion of typically central activities, but through their recentralization in the urban space, and it can also be associated with the formation of axes that unfold from the main center[10] on easy circulation routes, as well as the appearance of large commercial and service facilities such as *shopping malls*. In other words, the existence of several central areas in the urban space, in most cases, can be a consequence of the implantation of commercial and service equipment, such as *Prudenshopping* in Presidente Prudente, which causes, as Sposito (1998) pointed out, the redefinition of the center, the periphery and the center-periphery relationship.

Polynucleated centrality, on the other hand, refers to the tendency for new commercial and service facilities to be located in areas that are easily accessible (possibly sparsely populated) for social segments with high purchasing power. In other words, it means that the constitution of polynucleated centrality occurs not only from a functional point of view, but also from a socio-spatial point of view.

In addition to multiple centrality and polynucleated centrality, Sposito (2001a), Pereira

9 Central Business District.

10 Sposito (1991) makes an important contribution to the understanding of the main center. According to her, these are: "[...] characterized by the location of typically central activities, but in a specialized way, along roads with greater vehicular circulation. In addition, they are not areas contiguous to the main center or subcenters and therefore cannot be considered as their geographic expansion" (SPOSITO, 2001a, p. 243).

(2002) and Fernandes (2013) highlight **changing centrality**, which corresponds to that centrality that is continually redefined, varying over short time intervals, such as the pace of movement of vehicles and people, which vary differently in a given place throughout the day. This means that centrality is not only associated with what is fixed in space, but also with the relationships that are established in that space over time. According to Fernandes (2013, p. 34),

> [...] the **conditions of centrality vary over time, not just the long time that is counted in years or even centuries, but between summer and winter, weekends and other days of the week, as well as at various times of the day**, being seen and used as more central for some than for others, with clear differences between younger and younger people, visitors and residents, richer and poorer, more educated and less educated, among many dualities that simplify readings of the complexity and variability of each person, in a fragmented and diffuse centrality that can be seen on various scales. Many centers have the life of a city. [emphasis added].

Finally, Sposito (2013) highlights **planned centralities**, which are the result of the actions of agents who produce urban space, mainly developers and landowners, who act on urban space in a way in which actions are thought out long before they are lived.

The author also points out that the processes of segregation and fragmentation are, at the same time, contradictory,

> [...] processes that deny and redefine intra-urban centrality, as they transform it into a socially and functionally segmented centrality, dispersed in the territory and diffuse in the representation we make of the city itself and of the urban network, given that centrality can be understood and apprehended on multiple scales (SPOSITO, 2011, p. 138).

The emergence of new centralities in the urban space has been made possible by the widespread use of motor vehicles, which has led to changes in spatial practices, consumer habits and city dwellings, the emergence of housing estates in discontinuity with the urban fabric and in its peripheral spaces, as well as the movement of people with greater purchasing power towards gated communities on the outskirts, attracts the center and its offshoots to move in its direction (VILLAÇA, 2001).

In other words, the increase in the number of motorized families means that "[...] urban space is now conceived according to the pressures of the automobile, effectively one of the main factors responsible for redefining the places where people shop" (PINTAUDI and FRÙGOLI JR., 1992, p. 27) and live. On this issue, Sposito (2011, p. 139), based on Boudin (2005), points out:

> [...] the increase in mobility as an important element in understanding the trend towards individualization in society, dealing with this issue in terms of housing choices, consumption, leisure, sports activities, etc., in short, all kinds of socio-spatial practices that are part of the movement towards restructuring urban space.

With regard to the horizontal expansion of Presidente Prudente, an important issue highlighted by Sposito (1983) is that this expansion took place in a southerly direction, only when the psychological barriers were broken down[11] and the subsequent investment by the municipal and private authorities at the beginning of the 1980s. In this respect, it is worth remembering the important role played by the municipal authorities of Presidente Prudente in legalizing gated subdivisions, as highlighted by Sobarzo (2004, p. 102).

> In fact, the municipality's urban legislation, in force at the time of the implementation of the first two projects [Jardim Morumbi and Jardim Joao Paulo II], prohibited the closure of residential allotments. The municipal executive not only failed to comply with the legislation, but also contributed to the timely solution of the problem of illegality by sanctioning Municipal Law 2.187 in 1981 [...].In practice, this means that it is now easier to close down a subdivision. All that is required is to comply with the provisions of the zoning law, to have a maximum of two accesses, not to interfere with the extension of main roads, to have a guardhouse of at least 100 m^2 and to respect the indices that refer to the dimensioning of internal streets.

As well as serving as legal backing for the establishment of this type of enclosed residential space, this law contributed to the emergence of subsequent developments. Thus, in addition to the elements of the urban situation[12] , which was highlighted by Sposito (1983), the municipal government played an important role in the process of

11 Sposito (1983, p. 83): "We observe that for urban dwellers, railroads, highways or even rivers constitute not only geographical barriers, due to the difficulties caused for the circulation of cars and people, but also false psychological barriers, if we can call them that. It is common to hear people talk about areas located beyond the obstacles, such as "on the side of the river", "on the other side of the line", "after the road", etc."
12 It is defined by Pierre George (1983) as the location of the city in relation to natural facts that may, in the past or present, influence its development which, in turn, is linked to the ease of urban expansion.

expanding the urban fabric of Presidente Prudente towards the south/southwest.

According to Sobarzo (2004), in 2004 the city of Presidente Prudente had nine gated communities, of which only Residencial Portinari was located outside the south-southwestern part of the city. In 2014, the city now has fourteen[13] gated communities, three of which are located in the northwestern part of the city - Residencial Portinari, Porto Seguro Residencial *Resort* and Porto Bello *Residence*, while the others are located in the south-southeastern part. Remembering that "[...] these initiatives to establish closed residential areas, generating an exacerbated expansion of the urban fabric, induce and represent new ways of living and dwelling". (SPOSITO E GÓES, 2013, p.79).

Map 2 shows the proximity of closed residential areas[14] of high and medium occupational standard to car dealerships in Presidente Prudente, which can be explained mainly from two perspectives: (a) the existence of these walled residential spaces promotes the appreciation of adjacent areas (SPOSITO E GÓES, 2013) and; (b) being close to their potential consumers, who are the residents of the neighborhoods in the southern and western sectors of the city, as well as the avenues they usually use as a route, is a strategy adopted by the dealerships (c) both the condominiums and the car dealerships consider accessibility to be a fundamental condition in defining their locational choices.

The following three excerpts show different points of view on the relationship between the proximity of these closed residential spaces and car dealerships, in terms of their influence on the choice of location.

> It ends up automatically having [influence], because it's our audience, which favors us, it's an audience where we have a lot of customers of the brand... it's a customer profile that buys Peugeot, so it ends up influencing us as well. (Sales manager at *La Place Peugeot*).

> Proximity to residential neighborhoods does play a role, but according to the segment you operate in,

---

13 Considering only those of high and medium occupational standard, since if we were to consider the recent closed condominiums aimed at the low-income population, built with the help of the Federal Government's "Minha Casa Minha Vida" Program, this number would increase.

14 The use of the generic expression "closed residential spaces" is used in this text because some are de facto condominiums, while others are walled allotments that do not have the legal status of a condominium.

which is not my case. In my case it's a line, as I told you, it's a B segment, they're cars from 35 to 55 thousand reais, so this wouldn't be a decisive factor. (CAOA *Hyundai* salesman).

Not necessarily because the sales volume doesn't end up being standard. They are buyers of the more expensive cars, but it's not the standard of volume, of large numbers. So, I mean, the largest numbers are usually people from other neighborhoods and not from gated communities. (Autoeste *Chevrolet salesman*).

Directly related to enclosed residential spaces, we can see from map 3 that families with incomes above 20 minimum wages are mainly found in the central and south-southwestern portions of the city, often coinciding with these walled spaces.

Map 2

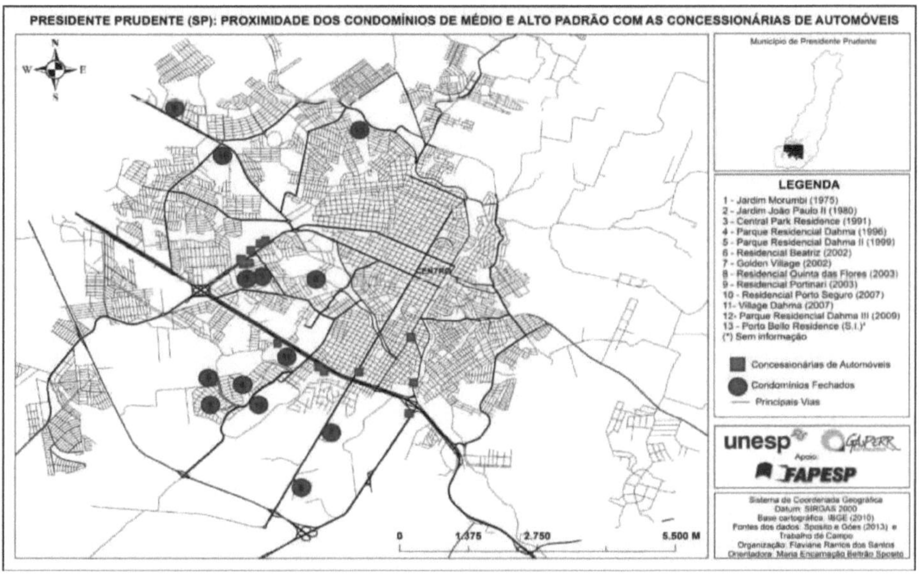

Source: Adapted from Sposito and Góes (2013) and fieldwork.

Map 3

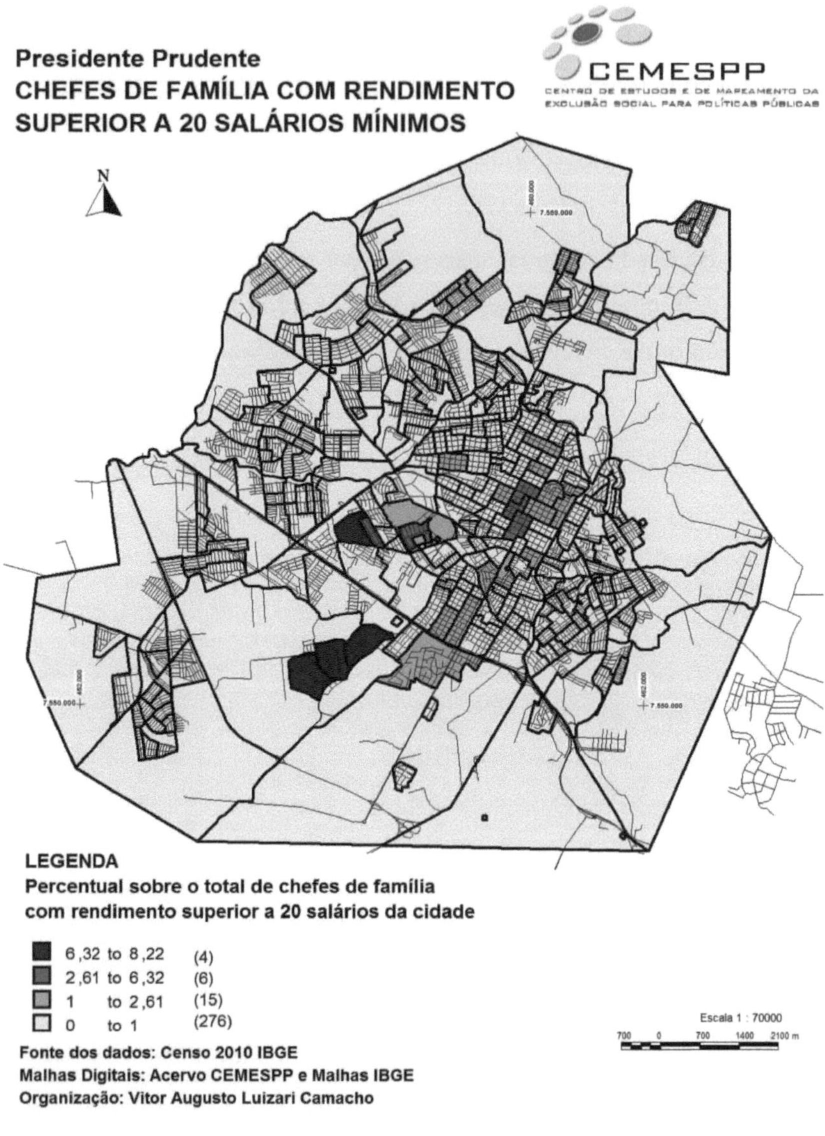

Source: CEMESPP, 2003[15]

In this way, the evaluation of the interview excerpts allows us to see that this proximity is much more related to the value that these gated communities give to their

15 The map legend is calculated from the distribution of occurrences (heads of household with an income of more than 20 minimum wages) in quartiles.

surroundings than it is to meeting the consumption demands of groups with greater purchasing power, mainly for two reasons: a) the vast majority of residents of gated communities already own a car and can therefore go to dealerships without too much trouble, regardless of whether they are nearby or not; b) although these wealthier residents buy more expensive cars, the bulk of sales come from other parts of the city, which are mainly looking to buy popular cars.

We would also point out that the higher incomes and the presence of this high- and medium-standard residential area in the south-southwest quadrant of Presidente Prudente is the result of the exacerbated appreciation of these new real estate products on the market, which are representative of the idea of security, even though crime rates are not as high as in the case of Presidente Prudente. At the same time, there is a redefinition of the center-periphery pair to the extent that:

> The distance between the unequal in the city is no longer predominantly based on the logic of the peripheralization of the poorest and the allocation to the richest of the central and peri-central areas, the best endowed with means of collective consumption (infrastructure, equipment and urban services). Urban security systems provide the conditions for this separation to deepen, even though social segments with unequal levels of purchasing power and different consumer interests are juxtaposed in the center and the periphery (SPOSITO, 2011, p. 140-41).

In addition, it can be understood through the concept of **segregation**, which according to Villaça (2001, p. 142) is "a process according to which different classes or social strata tend to concentrate more and more in different general regions or sets of neighborhoods [...]", however, segregation does not prevent the presence or even the growth of other classes in the same space.

More than the process and tendency towards segregation, we are witnessing the process of **socio-spatial fragmentation** in today's urban spaces, understood as "an advanced and more complex form of segregation, based on the growing appreciation of urban security in a world of imponderables" (SPOSITO, 2011, p. 124).

In this context, fragmentation differs from segregation in that, as well as producing new forms of urban *habitat*, clearly representative of the interest of separating or isolating the social segments with the highest purchasing power in urban spaces, it also generates a schism in the urban fabric, through the built walls and/or the discontinuity of the fabric resulting from the location of these new areas. However, according to Sobarzo (2004) and Sposito and Góes (2013), in relation to the structuring of the Prudentine urban space, it is still necessary to reflect on the idea of fragmentation, since spatial distancing via gated communities is much more a result of speculative economic interests than an actual response to the demand or desire for isolation.

Therefore, the concentration of car dealerships and other companies associated with commerce and services related to automotive transportation serves as an example of the process of restructuring the city and the valorization of the south-southwest quadrant, resulting from the occupation of "urban voids" created in the 1970s and 1980s, accompanied by the new "wave" of expansion of the urban fabric. In this way, the study of the spatial logic of companies that sell motor vehicles offers elements for understanding the links between **urban and city restructuring**. On the first scale, that of the urban network, it is important to note that it is companies representing the large international automakers that operate in the country and develop commercial and spatial logics according to the strategies and norms established by these transnationals. On a city scale, there have also been major changes, because while there used to be a tendency for dealerships to concentrate on the commercial and service axes that unfolded from the main center, more recently they have been established on more distant axes.

3. CAR DEALERSHIPS IN PRESIDENTE PRUDENTE

To begin this chapter, it is important to note that we have adopted the concept of Lima (2009), who defines a car dealership as a retailer of motor vehicles, recognized by specific legislation, which has exclusive rights to the respective vehicle manufacturer it represents, by means of a commercial concession contract.

According to Grande (2004), automakers sell their final product, i.e. the new car, to a network of numerous independent franchises that are responsible for selling the cars to the end consumer, which in turn are usually formed by a large local businessman, whose job it is to find the customer, close the deal and service the product sold.

In this way, what differentiates a dealership from another independent store is the so-called 'concession' of a particular brand, i.e. in order to represent *Ford, Fiat, Citroen* or any other, "a dealership must acquire the rights and duties laid down in the brand's 'flag', which provides for the exclusive marketing of products supplied by a single manufacturer" (LIMA, 2009, p.84), and must provide technical assistance and a warranty for all the vehicles in the range. The dealership can only sell products from a single supplier, and this supplier must guarantee the entrepreneur a quality and competitive product offer.

The concession contract for a new car dealership and its relationship with the car manufacturer is regulated by Law No. 6.729, of 28.12.1979, also known as the Ferrari Law, where the objective is clear and provides for the establishment of rules where the smaller economic power - that of the dealerships - has preserved its rights over the larger economic power - that of the car manufacturers (LIMA, 2009, p.84).

According to Lima (2009, p.84), the regulations are still very favorable to automakers, but "they do guarantee some rights for dealerships, such as preventing automakers from unilaterally terminating the contract".

The car dealership market is closely linked to car manufacturers, a direct relationship that is totally dependent on the conditions imposed by the car manufacturers, who dictate the rules of the partnership, such as: minimum vehicle purchase lots, fixed prices, service, structure and various other requirements (LIMA, 2009, p.85).

With regard to the size of companies that sell vehicles, Oliveira (2006) states that 45.5% of dealerships in Brazil are small family businesses, selling up to 40 units a month. The same author defines the concept of a family business based on three aspects:

> 1) **At the ownership level** - control of the company is in the hands of a family (which owns or controls the majority of the capital); 2) **At the management level** - the top positions in the company are occupied by family members; 3) **At the succession level** - the second generation of the family takes over the positions left vacant by relatives and so on (OLIVEIRA, 2006, p.17). - Emphases added.

In general, dealerships are physically organized with a *show-room* for displaying new vehicles, launches and vehicles for *test drives*, as well as their own or outsourced workshops and, in general, they are organized into various departments, which according to Lima (2009), are basically:

- Vehicle Department (new);
- Vehicle Department (used and used cars)
- Technical Assistance Department;
- Workshop and Parts Department;
- Bodywork and Painting Department;
- Financing and Insurance Department.

Faced with a highly competitive market, dealerships find themselves under pressure from customers on the one hand (who are looking for better prices and conditions) and from car manufacturers on the other (who control prices by publishing fixed prices). Therefore, in order to improve their profitability margins and as a way of differentiating themselves from the increasingly fierce competition, dealerships start offering additional services, such as dispatch services; insurance (complementary guarantees, assistance and help); and accessories (comfort items, safety and vehicle maintenance), among others.

With regard to the profile of car dealerships in Brazil, FENABRAVE (2014) points out that the vast majority are: a) small and medium-sized companies; b) have 100%

national capital and; c) are highly dispersed throughout the municipalities. Considering only ANFAVEA members, there are a total of 4,249 car dealerships in the country, representing 21 brands[16] (ANFAVEA, 2013), while in Presidente Prudente we have dealerships representing 13 brands (Graph 5), which will be detailed below.

Graph 5

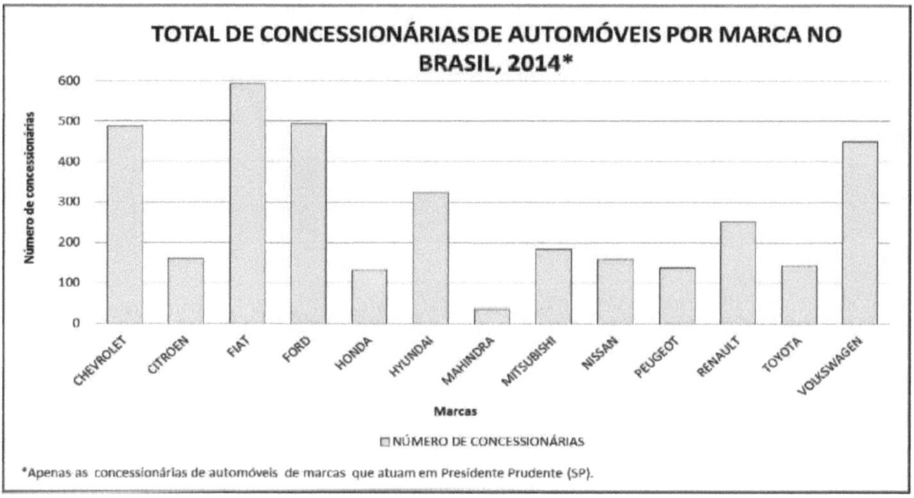

Source: ANFAVEA (2013) - Org. Flaviane Ramos dos Santos.

a) AUTOESTE CHEVROLET

The Autoeste *Chevrolet* dealership (Figure 2) belongs to a group of businessmen from the city of Catanduva (SP) and has been selling new and used vehicles, as well as parts, accessories and services in the city of Presidente Prudente since 1998. Its first location was at Avenida Antônio Canhetti, no. 201 (parallel to Avenida Manoel Goulart) and, since 2012, it has been located at Avenida Joaquim Constantino, no. 2871, on a site with a total area of 12,000 square meters, of which 5,000 are built. After the purchase of the land, the new building took around three years to be completed and brought up to the standard required by *Chevrolet*, according to the interviewee[17] .

16 The 21 brands that have car dealerships in Brazil that are members of ANFAVEA are: *Agrale, Citroen,* DAF, *Fiat, Ford, General Motors (Chevrolet), Honda, Hyundai, Iveco, Mahindra, Man, Mercedes Benz, Mitsubishi, Nissan, Peugeot, Renault, Scania, Toyota, Volkswagen* and *Volvo.*

17 All the interviews at the car dealerships located in Presidente Prudente were recorded with the permission

In 2013, Autoeste received the "A Classification", the highest award offered by *General Motors* do Brasil. This is because the Prudentine dealership achieved maximum points in the following areas: customer satisfaction, employee training index, facilities, sales performance, after-sales and financing.[18]

Figure 2 - Presidente PrudenteOoncessionaria Autoeste *Chevrolet.* 2014

Source: Fieldwork.

In addition to the city of Presidente Prudente, since 2003 Autoeste has had a dealership in Presidente Venceslau, located in the west of the state of São Paulo, approximately 60 km away from Presidente Prudente. According to the interviewee[19] , this was mainly to make it easier to serve customers from the states of Mato Grosso do Sul and Mato Grosso.

b) MITSUBISHI AURORA

One of the most traditional groups in Presidente Prudente, formed by three brothers and founded in 1967, the Aurora Group operates in different branches of activity. In addition to the Aurora *Mitsubishi* dealership (Figure 3), which began operations in 2003 at Av. Brasil, n°2107, the group also owns the Nova Aurora Maquinas Agricolas dealership and Aurora Retifica in Presidente Prudente and the Aurora Farm and the

---

of the interviewees, but not all of them gave their permission for their names to be disclosed. For this reason, we have chosen to withhold the names of all those interviewed.

18 According to news available at:< http://www.ifronteira.com/mobile/gente-53173>

19 Tables 1 and 2 in "4.2 Interviews and the vision of the companies" show the date and function of the interviewees.

Aurora Sugar and Alcohol Plant in the city of Anaurilândia in Mato Grosso do Sul, according to the interviewee.

Figure 3 - Presidente Prudente: Aurora Mitsubishi dealership. 2014

Source: Fieldwork

The building where the dealership is currently located was previously an agricultural implements workshop and the reuse of the space, which already belonged to the group, was the main reason for choosing the location, according to the interviewee. In 2012, the building was completely renovated to meet the standards required by *Mitsubishi*.

c) CAIADO FIAT/ VIVIANI FIAT

The Caiado Group, whose origins date back to 1954, is also one of the most traditional in Presidente Prudente and its history is linked, above all, to the automotive sector, since it began its activities by retreading tires, later moving on to selling truck tires and car tires, as well as workshop services. In March 2008, it acquired the *Fiat* dealership and opened Caiado Veiculos (Figure 4), a *Fiat* dealership in Presidente Prudente and Presidente Venceslau.

Figure 4 - Presidente Prudente: Caiado *Fiat* dealership. 2014

Source: Fieldwork.

Until 2007, the *Fiat* banner in Presidente Prudente belonged to the Ogata Group, based in Marilia. In 2008, the Caiado group bought and took over *Fiat*'s operations in the city and even changed its address, moving from Avenida Joaquim Constantino, 5555 to the extension of Avenida Manoel Goulart, 3636.

Despite being located in a rented building, Caiado has always invested in its facilities to provide greater comfort for its customers and employees, carrying out various renovations and extensions, including renting the neighboring buildings, where Alugue Brasil (a car rental company) and the Caiado Seminovos dealership, also belonging to the Group, operate, according to the interviewee.

In November of this year, the *Fiat* dealership in Presidente Prudente ceased to be the responsibility of the Caiado Group and was taken over by the Viviani Group, which has been based in Rio Claro since 1995. According to the interviewee, this was mainly due to a desire for growth on the part of the Viviani Group, but he was unable (or unwilling) to list the reasons for this, such as a possible lack of interest on the part of the Caiado Group in the Fiat dealership as a result of the fall in car sales and, consequently, in its profits, according to the following excerpt:

> Growth of the group. [Interviewer: Does this have anything to do with the Caiado Group's possible lack of interest in the dealership?] I don't know. But there was great interest from the group, which is a group that is growing, the car market in Brazil is also growing, so the group had this opportunity and that was it (Sales manager at Viviani *Fiat*).

Even with the change in management, the dealership has remained at the same address, at 3636 Avenida Manoel Goulart, with only the necessary changes to its façade (Figure 5). Despite being satisfied with the current location, the Group intends to relocate in the future (site not yet defined), with a view to operating in a larger building that better meets *Fiat*'s standards.

We're satisfied with the location, but we're interested in changing... trying to find a place with a much bigger front for the dealership... I think a Fiat dealership should have at least 40 meters of frontage and here we don't have that (Sales manager at Viviani *Fiat*).

The Viviani Group has *Fiat* dealerships in the cities of Rio Claro, Presidente Venceslau, Pirassununga and Porto Ferreira; *Toyota* dealerships in Araçatuba and Jales; and *Renault* dealerships in Bauru, Botucatu, Marilia and Jau.

Figure 5 - Presidente Prudente: *Fiat* Viviani dealership. 2014

Source: Fieldwork on 02/12/2014.

d) CAOA HYUNDAI

The CAOA *Hyundai* dealership (Figure 6) in Presidente Prudente is managed by the CAOA Group, the exclusive importer of the *Hyundai* brand in Brazil since 1999, which belongs to businessman Carlos Alberto de Oliveira Andrade .[20]

Figure 6 - Presidente Prudente: CAOA *Hyundai* dealership. 2014

20 Carlos Alberto de Oliveira Andrade is a doctor from Paraíba who owns dealerships representing the Hyundai, Ford and Subaru brands and a factory that produces for Hyundai in Anàpolis, in the interior of Goiás, according to the group's website.

Source: Fieldwork.

Before the CAOA Group took over *Hyundai*'s operations in the city in 2012 and moved to a rented building at Avenida Manoel Goulart, No. 3840, the brand's dealership had been run by *Special Motors* since 2010, which was located at Avenida Manoel Goulart, No. 3860, in the building next to where it is now, the reason for the move being physical space, according to the interviewee.

e) FORD V. MUCHIUTT

The Muchiutt family, traditionally known for operating in the hotel business and auto parts distribution, acquired the *Ford* dealership in the city of Presidente Prudente in 1998.

Located at Avenida Joaquim Constantino, 1895, the *Ford* V. Muchiutt dealership (Figure 7) not only sells new cars, but also after-sales services such as workshops, bodywork and paintwork and the sale of parts and accessories, as well as selling used cars of all makes in a building next door.

Figure 7 - Presidente Prudente: *Ford* dealership V.Muchiutt. 2014

Source: Fieldwork

Over the 16 years it has been operating in the city of Presidente Prudente, it has carried out several refurbishments and extensions, and will soon have a new expansion, given the increase in demand for vehicles and services provided by the dealership.

We're planning to extend it... it's *already* in the plans. There's another plot of land at the back, so we're going to extend it to make a division between light and heavy vehicles (Ford saleswoman V. Muchiutt)

In the city of Adamantina (SP), located 103 km away from Presidente Prudente, *Ford* V. *Muchiutt* has a branch.

f) LA PLACE PEUGEOT

The *La Place Peugeot* dealership (Figure 8), located at 1845 Avenida Joaquim Constantino, opened in 2006 in the city of Presidente Prudente. It belongs to the Pinezzi Group, based in Araçatuba (SP), which also has *La Place* dealerships in the cities of Araçatuba, since 2000, and in Marilia (SP), since 2011.

Figure 8 - Presidente Prudente: *La Place Peugeot* dealership. 2014

Source: Fieldwork.

According to the interviewee, by the end of 2014 *La Place* was planning to relocate to Presidente Prudente, leaving Avenida Joaquim Constantino and moving to the extension of Avenida Manoel Goulart, 3820, next to the CAOA *Hyundai* dealership. This move would be justified by the fact that, despite competing more directly with its main competitors, the increase in the flow of people and vehicles on Avenida Manoel Goulart is a significant aspect.

The refurbishment of the building on Avenida Manoel Goulart, which previously housed an evangelical church, lasted around two months and even had a sign on it saying "Future *La Place Peugeot* premises". However, as a result of a commercial disagreement between *La Place and the* owner of the building (and the person responsible for the renovation), after the renovation was finished this sign was replaced by a "For Rent" sign and the building is currently empty. According to the interviewee, with regard to the future, the dealership:

> Today, La Place is still studying what it is going to do, as there is a possibility that it will stay where it is today, or even move to another location, in the middle of Avenida Manoel Goulart, as it had hoped (Sales Manager at La Place Peugeot).

g) LIANE VOLKSWAGEN

The Liane Group, one of the most important groups in Presidente Prudente, began its investments with the acquisition of the Pasta Peretti noodle factory in the early 1960s, later renamed Liane Food Industry. Over time, the growth of the pasta industry allowed the group, in addition to investing in modernizing and expanding production, to diversify its investments into building materials stores, lumber yards, a transport company, a *Volkswagen* car dealership, a drinks distributor, a construction company and a property manager, forming one of the most important local economic groups. In recent years, with the downturn in the real estate market, the Liane Group has also been investing in beef cattle fattening in Mato Grosso do Sul, according to information on the Group's *website.*

The Liane *Volkswagen* dealership (Figure 9) was founded in 1985, at 1500 Antônio Rodrigues Street. With a large, modern structure, occupying a large part of the block, it offers sales and after-sales services (workshop, bodywork and paintwork, sale of parts and accessories), and in the future it plans to expand, with a view to occupying the entire block, which is already owned by the Group, according to the interviewee.

Figure 9

Presidente Prudente: Liane *Volkswagen* Dealership on Rua Antônio Rodrigues, 2014

Source: Fieldwork.

In 2007, two other Liane dealerships were opened in the city, one at Avenida Coronel Marcondes, 3018 (Figure 10) and the other at Rua Dr. Gurgel (Figure 11). However, unlike the main dealership, which has sales and after-sales services, these two only have a vehicle *showroom* (sales). In 2012, , the Liane dealership in the center closed down, according to the interviewee, because of the high rent.

Figure 10

Presidente Prudente: Liane *Volkswagen* dealership on Avenida Cel. Marcondes. 2014

Source: Fieldwork.

Figure 11

Liane Presidente Prudente: Vehicles on Rua Dr. Gurgel, in the main city center. 2012

Source: Image from *Google Street View* (June 2012).

It's important to note that the Liane *Volkswagen* dealership, until then, was the only one that had opened branches in the city, a fact that can be understood by the peripheral location of its head office, Rua Antônio Rodrigues, n° 1500. In other words, with the aim of giving it more visibility and attracting more consumers, and thus increasing its profits, the Liane *Volkswagen* dealership chose other important locations in Presidente Prudente, where the daily flow of people is very significant, to place stores with *showrooms* for its vehicles. As will be seen below, Via *Japan Toyota,* which also has a peripheral location, also has a branch in the city.

h) LIFAN/MAHINDRA KOREMOTOR

The *Koremotor* dealership (Figure 12), *an* authorized distributor of the *Mahindra* brand, arrived in Presidente Prudente in 2009 through the Pinezzi Group, which also manages *La Place Peugeot* in the city. In 2012, the group sold the dealership to a local businessman, but it remained at the same address, at Avenida Coronel José Soares Marcondes, 2907 near Parque do Povo.

Figure 12 - Presidente Prudente: *Lifan and Mahindra Koremotor* dealership. 2014

Source: Fieldwork.

At the beginning of 2014, he acquired the *Lifan* dealership and moved to a rented building at Avenida Antônio Canhetti, No. 247. The reason for the move, according to the interviewee, was physical space, i.e. the need for a bigger space, since the dealership now sells cars from two brands (*Lifan* and *Mahindra*), as well as multi-brand used cars. Another issue is that the dealership now offers after-sales services, unlike in the old building, which didn't have enough space and the service was outsourced.

Two other advantages cited by the interviewee of the new location compared to the previous one are the ease of parking and the proximity to the city exits, since its main consumers are rural landowners, as Mahindra cars are typically utilitarian.

i) CITROEN LIGHT

The *Lumière* Group, headquartered in Sao José do Rio Preto (SP), is responsible for managing the *Citroen* dealership in Presidente Prudente, which opened in 2005 at Avenida Washington Luiz, 2375. In August 2009, *Lumière* moved to Avenida Antônio Canhetti, 319 (Figure 13), because of the need for a larger building, given the increased demand for *Citroen* cars in the city, as well as for services.

Figure 13 - Presidente Prudente: *Lumière Citroen* dealership. 2014

Source: Fieldwork.

In the building next to *Lumière*, there was the dealership that represented the *Cherry* brand, which, with just under a year of operation, left Presidente Prudente in mid-2011, leaving the entire structure of the building in the standards of a dealership. So, in order to take advantage of this structure and the proximity, *Lumière* opened its multi-brand dealership, where cars acquired in exchange were sold, i.e. when a customer offers their used car as part payment for the purchase of their *Citroen*. However, in mid-August 2014, *Lumière* closed its multi-brand dealership and these cars were once again sold at the main dealership, along with new vehicles.

In addition to Presidente Prudente, the São Paulo cities of Bauru, Marilia and Araçatuba have *Lumière Citroen* dealerships, as does Sao José do Rio Preto.

j) VIA JAPAN TOYOTA

The *Via Japan Toyota* dealership (Figure 14) arrived in Presidente Prudente in 1998 and set up shop at Avenida Antônio Canhetti, No. 321. Recently, in 2009, the Enzo

Group[21] from Campo Grande in Mato Grosso do Sul acquired the *Toyota* dealership in Prudente and, due to the need for expansion, rented and completely renovated a building located on Rodovia Assis Chateaubrand, km 454, near the junction with Rodovia Raposo Tavares.

Figure 14 Presidente Prudente: Via *Japan Toyota* dealership on the Assis Chateaubrand highway. 2014

Source: Fieldwork.

Like Liane *Volkswagen*, in November 2014 Via *Japan Toyota* opened a branch in the city (Figure 15), at Avenida Manoel Goulart, 3636 B, next to *Fiat* Viviani and where the Caiado Seminovos dealership used to be. Although it is called Via *Japan* Seminovos, the dealership has new *Toyota* cars on display and even in a prominent position, while the multi-brand and used cars are inside the store. According to an interview with a sales assistant at the Via *Japan* branch:

The main reason [for opening a new *Via Japan* unit in the city] is the point. Here we are inside the city, there we are at the exit, at the end, already going to Pirapozinho. So I think the biggest reason is the location. [Interviewer: What advantages does Manoel Goulart have over the highway? Flow of people, mainly. There, for example, the person passing by is either going on a trip or is very interested in the car and is going there because they really want to buy. Not here, we take the people who are passing by and who see the cars every day (Via *Japan Toyota* salesman).

Figure 15

21 The Enzo Group is the largest in the automotive sector in Campo Grande (MS), where it represents the *Toyota, Jaguar, Peugeot, Fiat, Volare, Agrale, Mercedes-Benz Trucks* and *Harley-Davidson* brands.

Presidente Prudente: Via *Japan* Seminovos dealership on Avenida Manoel Goulart. 2014

Source: Fieldwork, 02/12/2014.

Again, as in the case of Liane *Volkswagen*, the justification for this new store in the city is to increase the number of consumers, because although *Via Japan,* located on the highway, has enormous visibility, accessibility to it is relatively poor, since once you enter the Assis Chateaubrand highway towards the city of Pirapozinho, you have to travel around two kilometers to return to the city of Presidente Prudente.

[Interviewer: Is the feedback you get when you leave Via Japan on Assis Chateaubrand unfavorable?] It's unfavorable, because the customer wants it easy, so they want to buy and leave quickly. There, for example, the customer has to go to Via Japan, which is already far away, and then they have to walk another 2 km to get back to Prudente, so it's far away and not very favorable. (Via *Japan Toyota sales* assistant).

k) VILLE NISSAN

The *Ville Nissan* dealership belongs to the *Ville* Group of Sao José do Rio Preto, which also manages *Ville Renault* in Presidente Prudente. It has been located in the city since 2007, and its first address was at 4000 Manoel Goulart Avenue, where it shared a building with the *Renault* dealership. In 2011 it moved to number 3700 on the same avenue (Figure 16).

Figure 16 - Presidente Prudente: *Ville Nissan* dealership. 2014

Source: Fieldwork.

According to the interviewee, the group's choice to open the two dealerships in the same space was due to the fact that, at first, they wanted to check the acceptance of the respective brands in the Prudentine market. In 2011, after seeing the success in sales, Nissan moved to Avenida Manoel Goulart, 3700.

My boss acquired the two concessions, Renault and Nissan, and at the beginning we didn't know how the Renault and Nissan cars would be accepted by the market here in Prudente, so we put them in the same building. But it was a success, thank God, then the range of cars increased and Nissan developed very well, so we had to move the building because it became too small for us to work in. So, to be a bit closer, he put it here on the same street to be really close and broke up Nissan, but it's still the same group and the same owner. It was just a question of space (Ville Nissan saleswoman).

The cities of São José do Rio Preto, Votuporanga and Araçatuba also have *Ville Nissan* dealerships.

l) VILLE RENAULT

In March 2007, *Ville* Veiculos, an authorized dealership for the *Renault* and *Nissan* brands, opened on Avenida Manoel Goulart, n° 4000 in Presidente Prudente, and in 2011 *Nissan* moved to another building on the same avenue, as explained in the previous topic (Figure 17).

Figure 17 - Presidente Prudente: *Ville Renault* dealership. 2014

Source: Fieldwork.

In addition to Presidente Prudente, the cities of Sao José do Rio Preto, Araçatuba, Votuporanga and Catanduva have *Ville Renault* dealerships belonging to the same group.

m) V.M. CENTER HONDA

The V.M. *Center Honda* dealership (Figure 18), inaugurated in 2005 in the city of Presidente Prudente, is located at Avenida Coronel José Soares Marcondes, 3911, with an 800 square meter *showroom* and large areas for technical assistance services (workshop and body shop). In addition, V.M. *Center* is another of the businesses that the Muchiutt family owns in the city, as well as the *Ford* dealership, a hotel and an auto parts distributor.

Figure 18 - Presidente Prudente: V.M. *Center Honda* dealership. 2014

Source: Fieldwork.

Over the nine years it has been operating in the city, three refurbishments have been carried out. At first, the adjacent buildings were rented to increase the physical space, and then other investments were made in order to have a modern *showroom* for displaying the brand's vehicles. According to the interviewee, the dealership's future plans are:

> Expansion, for sure. Relocation and expansion. [Interviewer: Where are you thinking of moving from here? Do you have any ideas?] To the highway itself, to the side of the highway (After-sales manager at V.M. Center).

In the city of Assis (SP), 128 km from Presidente Prudente, there is another V.M. *Center Honda* dealership.

4. STRUCTURING THE URBAN SPACE AND THE SPACE LOGICS OF CAR SALES COMPANIES

Population and territorial growth, as well as the city's economic development, do not happen uniformly, either in time or in space. In this way, as Corrêa (1989) points out, we must consider that the city is historically produced by different actors who often act in a contradictory way, because they have different objectives and power to act, in terms of access to the best locations and their implementation over time.

In the case of the city of Presidente Prudente, its population has grown significantly over the last 40 years, doubling from around 105,000 inhabitants in 1970 to almost 208,000 in 2010, as shown in Table 1:

Table 1

Municipality of Presidente Prudente - Population growth from 1970 to 2010

Year	Urban Pop. Urban (%)	Total Population
1970	88.2%	105.107
1980	94.7%	136.943
1991	96.8%	165. 484
2000	96.2%	189.186
2010	97.9%	207.625

Sources: IBGE Demographic Censuses of 1970, 1980, 1991, 2000 and 2010.

The city's rapid growth from the mid-1970s onwards, according to Sposito (1983), was a reflection of the transformations that took place in the countryside, such as the replacement of agricultural products by livestock, as well as the migration of people from smaller towns who, in search of better living conditions and employment, migrated to Presidente Prudente. This dynamic explains why this city and its commercial and service functions have grown, generating a relative loss of importance for the tertiary sector in the small towns it commands.

Underlying this increase in the city's population is its spatial growth, which has caused changes in its structure, especially since the second half of the 1970s, as mentioned above, since demographic growth, although not the only determinant, has also been an inducer of the city's territorial expansion, through the opening of allotments on the outskirts, often leaving extensive urban voids for future appreciation (SPOSITO, 1983).

In order to understand the logic of urban territorial expansion in Presidente Prudente from the 1970s onwards, Sposito (1983) provides us with a rich contribution. She states that the city experienced unprecedented urban expansion, especially from 1976 onwards, without, however, this expansion of the urban network through the opening of allotments being directly accompanied by an increase in the city's population. Thus, the author found that "[...] land sales grew much faster in the 1970s than could be considered necessary [...]" (SPOSITO, 1983, p.81).

In the 1980s and 1990s, as a result of demographic growth and speculative logics, and in the 2000s, as a result of these dynamics and the increase in purchasing power in Brazil - which was reflected in the city - there was an increase in the price of available land and a stimulus to the parceling out of these "urban voids", at the same time as the city continued to expand discontinuously, generating other areas of this type (SPOSITO, 1983).

This dynamic leads to a relocation of commercial and service activities in the urban space, generating functional changes in the use of urban land as well as changes in its social content. In other words, we can see a restructuring of the city through the location of commercial and service activities that have begun to emerge or relocate as a result of the growing segmentation of residential use of urban space.

As already explained in Chapter 2, which deals with the production of urban space by economic agents, a good example of this restructuring of the city is the increase in value of the south-southwest quadrant, due to the occupation of "urban voids" created in the 1970s and 1980s, accompanied by the new "wave" of expansion of the urban fabric. This can also be explained by the geographical location of this part of the city, where the topography is made up of long, gently sloping hills, which offer fewer difficulties in terms of topography compared to the eastern part.

As a result, car dealerships are concentrated mainly along Avenida Manoel Goulart[22] and Avenida Antônio Canhetti and Avenida Joaquim Constantino in Presidente Prudente, in the south-southwest sector. The first avenue is home to Viviani *Fiat*, CAOA *Hyundai, Ville Nissan and Ville Renault;* the second avenue is home to *Lumière Citroën and Koremotor (Mahindra and Lifan)*, while the third avenue is home to *Ford* V. Muchiutt, *La Place Peugeot and Ville Renault. Muchiutt, La Place Peugeot,* Autoeste *Chevrolet* and Via *Japan Toyota*. The three exceptions are the V.M. *Center Honda*, Aurora *Mitsubishi* and Liane *Volkswagen* dealerships, which, although they are not on the three aforementioned axes, are very close to one of them, Joaquim Constantino, according to map 4.

Mapa 4

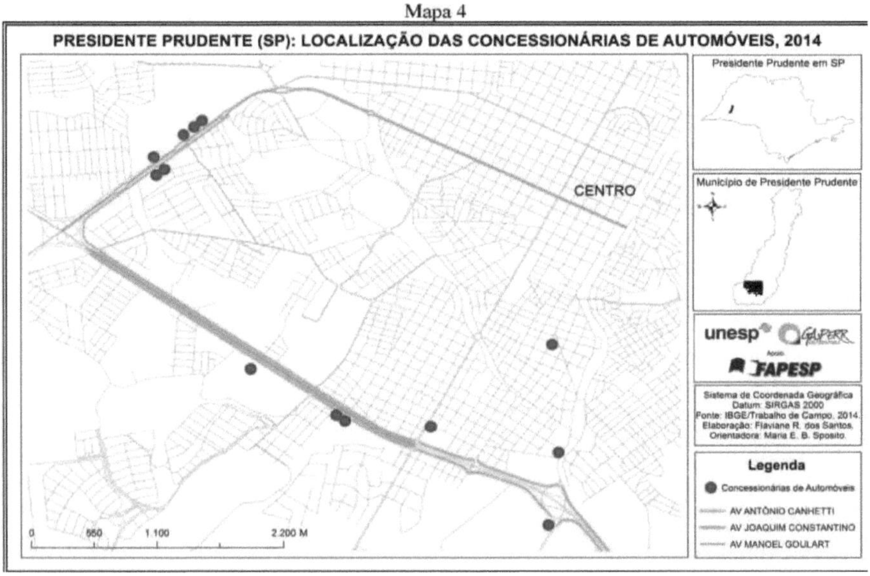

Source: IBGE (2010) and Fieldwork.

Regarding the location of car dealerships, it is important to remember that there is no impediment on the part of the public authorities, in accordance with the zoning established by law in the city's master plan, with regard to the use and occupation of urban land, in relation to this branch of commercial activity. Concessionaires have total

22 Remember that in this study we only considered the extension of Avenida Manoel Goulart, i.e. the stretch between *Prudenshopping* and Rodovia Raposo Tavares.

freedom to choose the part of the city in which they want to set up, according to their interests, i.e. in this case, the interests of private initiative and property owners dominate the spatial choices of entrepreneurs, with no restrictions on the part of the public authorities.

In addition to the presence of car dealerships, we found, through a survey of all the commercial establishments located on these avenues (from the 2013/2014 Telephone Directory), a large number of companies associated with commerce and services focused on automotive transport, such as auto parts stores, fuel stations, tire and auto parts distributors, among others, as shown in Graph 6.

Graph 6

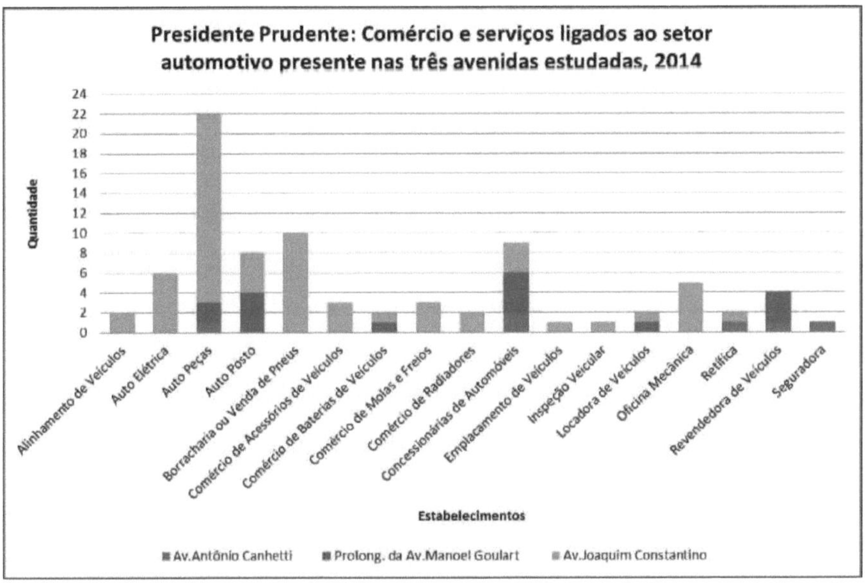

Source: 2013/2014 telephone directory and fieldwork.

According to the graph, there is a greater occurrence of these establishments on Avenida Joaquim Constantino than on the extension of Avenida Manoel Goulart or Avenida Antônio Canhetti, which is mainly due to the difference in the length of these avenues.

4.1. Location of car sales establishments in Presidente Prudente: the origin associated with the main center

To draw up a historical retrospective of the location of establishments specializing in car sales in Presidente Prudente over the last 20 years (Map 5), we used information from the Presidente Prudente telephone directories for 1988, 1998 and 2008. Our intention, in this case, was to gather elements to analyze the changes in the locational strategies of entrepreneurs, taking as a reference the fundamentals of the bibliography we considered, according to which the city is produced by social agents with different and often contradictory objectives and powers of action.

This source does not differentiate between a dealership (which has a concession contract with a brand) and a car dealer (which only sells used and multi-brand cars, without, however, having acquired the concession to do so). Therefore, in this subchapter only, we will use the term 'car dealer' to refer to these establishments specializing in the sale of motor vehicles.

Returning to what has already been mentioned, the locational choices of car dealerships and, in the case of this topic, car dealerships, are the sole responsibility of the dealers, since the government has no restrictions on their location, as is the case with other types of establishments, for example. In the absence of any restrictions, the desire to be close to other establishments that also sell cars, or the choice of commercial hubs that provide ample accessibility and visibility are important strategies adopted by these agents.

Another issue that deserves to be highlighted is the role of the automobile in changing the spatial practices and consumption habits of city dwellers, since the emergence of new centralities in the urban space has been made possible by the widespread use of motor vehicles. In other words, the choice of dealers to move away from the main center, as shown in map 5, is directly related to the improvement in the mobility of city dwellers, which has, in a way, relativized distances.

Mapa 5

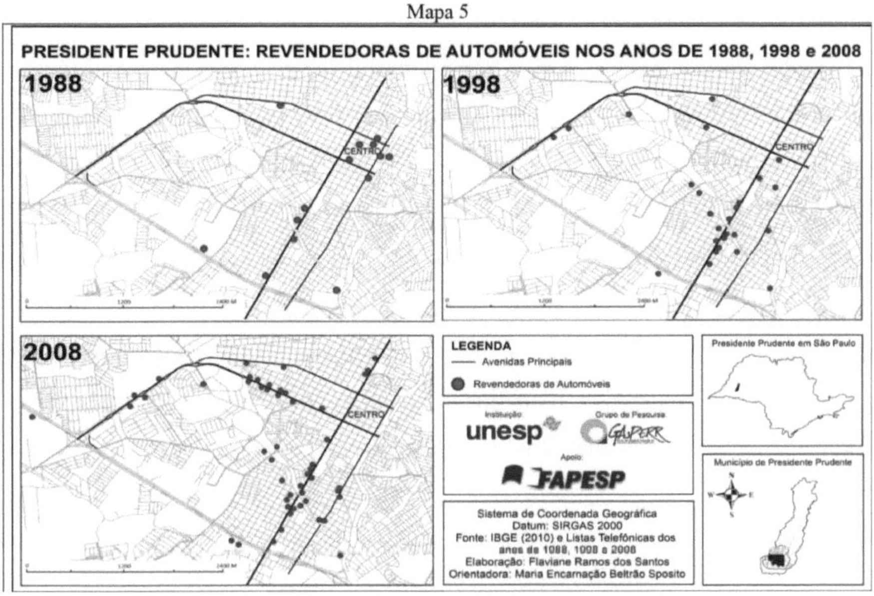

Source: IBGE (2010) and telephone directories from 1988, 1998 and 2008.

At the end of the 1980s, we can see that the car dealerships in Presidente Prudente were mainly located in the central portion and at the end of Avenida Coronel José Soares Marcondes, i.e. on one of the city's main entrance roads.

This concentration in the center can be explained by the fact that, in the 1980s, this was the most accessible sector and where all the city's flows converged. In addition, it was the area that coordinated the already consolidated urban fabric, with urban infrastructure, facilities and services.

At the end of the 1990s, more specifically in 1998, car dealerships were no longer located in the main center of Presidente Prudente. They were dispersed throughout the city, due to new locational logics, although they still had a significant concentration on Avenida Coronel José Soares Marcondes, near Avenida Quatorze de Setembro and Avenida Onze de Maio, in other words, in Parque do Povo.

The departure from the center may have occurred for three main reasons. Firstly, the increase in the value of land located in the center and the consequent rise in prices. Secondly, the need to move away from the main center in order to avoid parking difficulties for consumers, which implicitly reveals the clientele primarily targeted and served, as well as parking and traffic difficulties for the shame trucks. And thirdly, because of the emergence of new centralities, i.e. "urban organization has therefore moved from a situation dominated by one center to one of polycentrism in which there are several areas of centrality, some of which are specialized and others which are not [...]" (BARATA-SALGUEIRO, 2013, p.8).

According to Villaça (2001), the emergence of new centralities is due to the increase in spatial mobility, a consequence of the rise in the rate of motorization of the population, which has expanded the radius of action in which people can make use of commerce and services, as well as leisure and residence.

> [...] the spread of car use, together with the commitment of real estate capital to make existing centers obsolete and promote new ones, has led to a violent process of decentralization that has affected almost all Brazilian metropolises and even medium-sized cities (VILLAÇA, 2001, p.281).

In 2008, the number of car dealerships increased significantly and the most interesting thing is that these new establishments tried to set up near other established dealerships, resulting in the emergence of specialized commercial hubs, such as Avenida Coronel José Soares Marcondes, near Avenues Quatorze de Setembro and Onze de Maio (Parque do Povo) and Avenida Manoel Goulart, on the stretch between Avenues Quatorze de Setembro and Celestino Figueiredo.

In this way, we can see that there has been a change in the spatial logics practiced by economic agents in the car sales business, since their first establishments sought to be located in the main center of the city, since this was the place where the flow of people was most intense. Nowadays, however, this strategy no longer corresponds to consumer needs, since the city has expanded horizontally and the automobile has facilitated the physical mobility of consumers and, as a result, distances have become shorter.

In short, it is possible to observe that the number of these establishments has increased in recent years, above all by attracting brands that used to be present only in the metropolises. We also found that they have dispersed throughout the city over the 20 years analyzed, due to new locational logics, and that this fact is directly related to the increase in the car fleet in the city of Presidente Prudente, as shown in Chapter 1.

4.2. The interviews and the companies' vision

In recent years, interviews have been widely used by researchers who need to obtain data that cannot be found in records and documentary sources, but which can be provided by well-informed agents. In the case of this study, we used interviews as an instrument for gathering information and data on the proposed problem. To this end, three documents were drawn up: a letter of introduction, an interview form and an interview script.

The introductory letter, which was handed out before the interview, was intended to briefly explain the objectives of the research and make it clear that it was strictly academic, as well as guaranteeing the confidentiality of the information provided.

As for the interview form, it was based on the model provided by GAsPERR, which is widely used in other research on the Thematic Project with which this project is associated , and then made the necessary adjustments so that it met the needs of this work.

Finally, a script was drawn up containing 26 questions divided into three blocks. The aim of the questions was to understand the dealerships' strategies in the urban space, the main factors they attribute to the location of their establishment, such as proximity to city exits and high-income neighborhoods, distance from the main center and large flows of people and vehicles, among others, as well as the possible advantages or disadvantages of proximity to competing dealerships.

Interviews were carried out at all the dealerships, and the order in which they were carried out was defined according to their location, i.e. first, those along Avenida Manoel Goulart and Avenida Antônio Canhetti, then those located on Avenida Joaquim Constantino and its surroundings (such as Liane, *Via Japan* and VM *Center*) and, lastly, *Mahindra Mumbai Motors*[23] and Aurora *Mitsubishi*, far from the others.

23 The *Mahindra* dealership was still located on Avenida Coronel José Soares Marcondes on the day of the interview.

Table 1 shows the date and order of the interviews, as well as the position held by the interviewee in the respective company.

Chart 1 - Presidente Prudente: Car dealerships. Interviews conducted, 2013

DATE	CONCESSION	INTERVIEWEE
Nov. 13, 2013	Ville Renault	Saleswoman
Nov. 13, 2013	Caiado Fiat	Seller
Nov. 13, 2013	Ville Nissan	Seller
Nov. 13, 2013	CAOA Hyundai	Sales Manager
Nov. 13, 2013	Lumière Citroen	Seller
Dec. 16, 2013	Liane Volkswagen	After Sales Manager
Dec. 16, 2013	La Place Pegeout	Sales Manager
Dec. 16, 2013	Autoeste Chevrolet	Seller
Dec. 16, 2013	Ford V. Muchiutti	Administrative assistant
Dec. 16, 2013	V.M. Center Honda	Sales Manager
Dec. 16, 2013	Via Japan Toyota	Seller
Dec. 17, 2013	Aurora Mitsubishi	Sales Manager
Dec. 17, 2013	Mahindra/Lifan Koremotor	Seller

Org. Flaviane Ramos dos Santos.

These interviews provided valuable input for the proposed study, as it was possible to identify important issues, such as the year in which they set up in the city and, in some cases, the year in which they moved, as well as their opinion on proximity to other dealerships, average number of employees, among other information.

In view of some of the changes that took place in November 2014, new interviews were carried out at Caiado *Fiat*, Via *Japan Toyota* and *La Place Peugeot*, in order to understand the reasons behind the sale of the *Fiat* dealership in Presidente Prudente, the opening of a Via *Japan* branch in the city, as well as *La Place*'s decision to move to the extension of Avenida Manoel Goulart, among other things already described in Chapter 3. Table 2 shows the date and position held by the interviewee at the respective dealership.

Chart 2 - Presidente Prudente: Car dealerships. Interviews conducted, 2014

DATE	CONCESSION	INTERVIEWEE
December 16th, 2014	La Place Peugeot	Sales Manager
December 16th, 2014	Via Japan Toyota Seminovos	Seller
December 16th, 2014	Viviani Fiat	Administrative Manager

Org. Flaviane Ramos dos Santos.

With regard to the year in which the dealerships set up in Presidente Prudente, we can see that this occurred mainly during the 2000s, i.e. of the thirteen dealerships in the city, nine arrived between 2003 and 2010. Of the others, three arrived between 1998 and 1999 and only one in 1983, which is the oldest in the city, according to Graph 7.

Graph 7

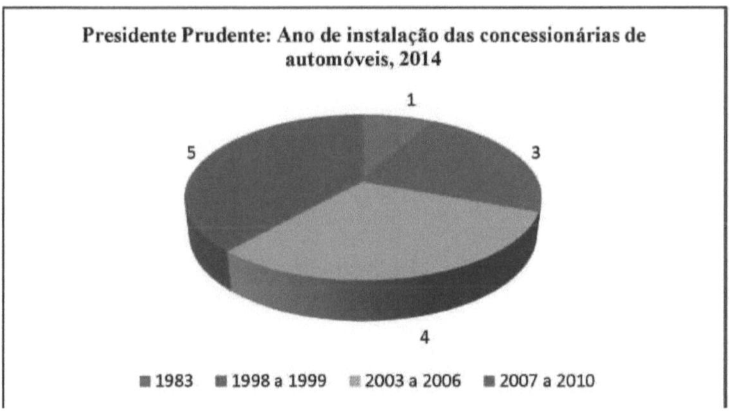

* Determining classes by reciprocal pairs. Source: Interviews

When we map the car dealerships according to the year they arrived in Presidente Prudente (Map 6), we get a very interesting result, as we can see that, above all, those that have been in the city the longest (those that set up between 1983 and 1999) are located on Avenida Joaquim Constantino or, in the case of Liane, very close to it, such as V. Muchiutt, Autoeste and *Via Japan*.

The four that arrived in the city between 2003 and 2006 are each on a different avenue: Aurora is on Avenida Brasil, a long way from the other dealerships; V. M. *Center* is on Avenida Coronel José Soares Marcondes, very close to Avenida Joaquim Constantino; *Lumière* is on the extension of Manoel Goulart; and *La Place is* on Joaquim Constantino.

On the other hand, the most recent dealerships, set up between 2007 and 2014, such as Viviani Fiat, CAOA Hyundai, *Ville Renault*, *Ville Nissan* and *Koremotor*, chose the extension of Avenida Manoel Goulart as the place to set up their establishments, indicating a recent trend towards the functional specialization of this road in the motor vehicle trade.

Map 6

PRESIDENTE PRUDENTE (SP): ANO DE INSTALAÇÃO DAS ATUAIS CONCESSIONÁRIAS DE AUTOMÓVEIS

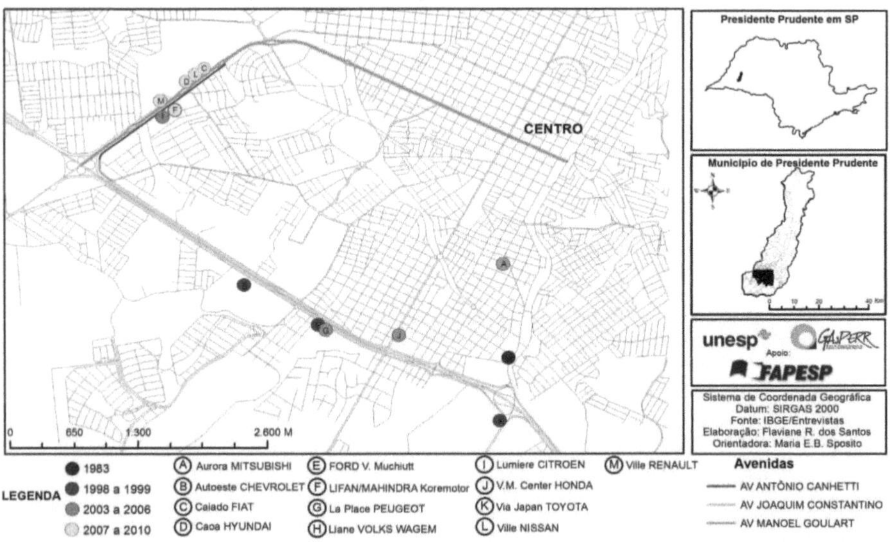

Source: IBGE (2010) and interviews.

Regarding their choice of location, the main factors mentioned by the interviewees of the dealerships located on the Manoel Goulart Avenue and Antônio Canhetti Avenue extensions were, above all, the importance of the flow of people and vehicles, landmarks such as the Christ the Redeemer statue and *Prudenshopping*, as well as the presence of competitors as determining factors in their choice of location, according to the excerpts below:

In fact, most dealerships choose where they are concentrated and today Avenida Manoel Goulart, from what I understand, has the largest number of dealerships, so it's a strategic point. It's an avenue with easy access, it's a flow avenue. Proximity to the city's exits is also important because, whether we like it or not, the stagecoaches can't travel within the big city center and it's also an easy route for those coming from the region, so it favors us in terms of sales, right? (Sales Manager at CAOA Hyundai).

The dealership business is more located in this area. There's a mall nearby, there's a flow... it's easier to locate, because when someone says: "Where is the dealership near?", we say: "It's near Christ", so it's very practical (Nissan sales assistant).

The agents of companies located on or near Avenida Joaquim Constantino highlighted three factors in particular: accessibility, both from the point of view of consumers coming from other cities, and from the point of view of stork truck traffic; the availability of land and; the visibility provided by the location on the edge of the highway, as shown in the following excerpts:

Easy access for customers... The highway, both for those who are in the city and for those who come from outside, knowing that Prudente is a hub, all the cities in the region end up coming here, and also the ease of transporting cars, because it's difficult for trucks to enter the city to unload. Here it's easier, because the truck comes, unloads, leaves and there's no headache, and it doesn't get in the way of traffic in the city (Autoeste Chevrolet salesman).

Two factors. The first is access and the view from the side of the highway, hence the marginal road, which provides very easy access both for customers and also for unloading cars and trailers. And the other factor is also the cost of the property in terms of rent (Sales Manager at Peugeot).

Actually, space. We needed space, Toyota itself demanded more space and that's why. And in fact the location favors us on one point and disadvantages us on another, for example, we have to make a journey... the customer has to turn around down there to leave [referring to the 2km journey to return to the city] and this makes it difficult and we see that customers complain about this. But the location is good for us, because of the comfort we give the customer, the space inside for parking, the unique visibility, the customer comes here and we are unique, where we were there were several next to each other, so you were selling one car, but the customer is already seeing the other (Sales assistant at Via

Japan Toyota).

Other factors also cited were the low cost of the property compared to other locations, large physical space, easy parking and the building itself, in some cases.

The land already belonged to Mr. Valter, so I won't be able to give you a precise answer, because he's had it since 2001. So it was a piece of land that was sitting idle, he actually built this building to rent out. In 2004 he received an invitation from Honda to set up a dealership in Prudente, and he already had the building ready, so he ended up using the same building. [Interviewer: And what benefits does this address bring you? Good location, close to the highway, easy to get to by stork... so that's basically it (Sales Manager V.M. Center Honda).

It was already ours. Here was a workshop for... of chains for deforestation, which was agricultural implements, it was a factory... very old, very old, so we took advantage of the space (Sales Manager at Aurora Mitsubishi).

It's important to note once again that some of these dealerships have already changed their address (Graph 8), as is the case with *Via Japan Toyota* and Autoeste *Chevrolet*, which were previously on the extension of Avenida Manoel Goulart and moved to Avenida Joaquim Constantino in 2009 and 2012, respectively. In addition, *Lumière Citroen*, which used to be on Avenida Washington Luiz, and *Mahindra Koremotor*, which used to be on Avenida Coronel José Soares Marcondes, moved to Avenida Antônio Canhetti. The Caiado *Fiat* and CAOA *Hyundai* dealerships have also changed address, with the former moving from Avenida Joaquim Constantino to the Manoel Goulart extension and the latter moving only to the neighboring building, remaining on the same avenue.

Graph 8

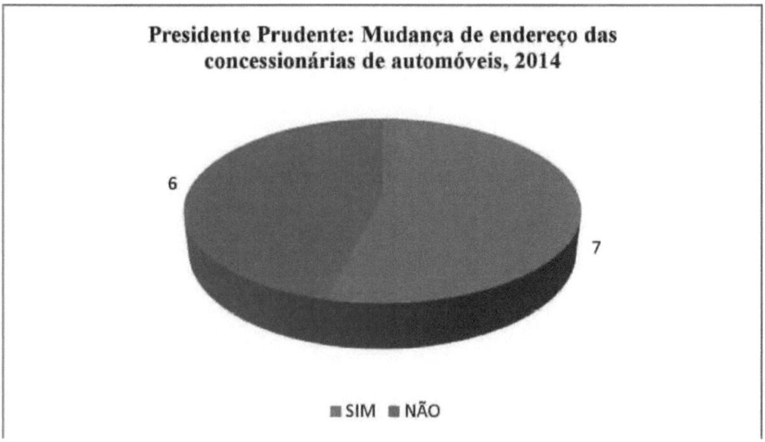

Source: Interviews.

According to the data in Table 2, we can see which car dealerships in Presidente Prudente have never changed their address, as well as those that have.

Table 2 - Presidente Prudente: Year of arrival and change of address of car dealerships, 2014

DEALERSHIPS	YEAR OF ARRIVAL IN THE CITY	YEAR OF CHANGE OF ADDRESS
Aurora Mitsubishi	2003	
Autoeste Chevrolet	1999	2012
Caiado Fiat	2008	2008
Caoa Hyundai	2010	2010
Ford V. Muchiutti	1998	
La Place Peugeot	2006	
Liane Volkswagen	1983	
Lumière Citroen	2005	2009
LifanMahindra Koremotor	2009	2014
Via Japan Toyota	1998	2009
Ville Nissan	2007	2011
Ville Renault	2007	
V.M. Center Honda	2005	

Source: Interviews - Flaviane Ramos dos Santos.

These relocations to urban areas are mainly due to the need for more physical space. In other words, in order to meet the automakers' requirements, dealerships are renovating and expanding their physical spaces, and when this is not possible, some of them are choosing to relocate to urban areas, especially in places where adjacent areas are available for further expansion in the future.

Map 7 shows exactly this dynamic: of the thirteen car dealerships located in Presidente Prudente, seven have already relocated to the city and, according to the reports, the main reason was precisely the need for a larger physical space.

At this current address, the advantages are that the building is bigger, it's 2,000 meters longer than the previous one, it's easier for customers to park and the location is good too, it's on an avenue with a lot of dealerships (Saleswoman at Lumière Citroen).

As a result, three dealerships moved to the extension of Avenida Manoel Goulart and Avenida Antônio Canhetti, while two only moved buildings and stayed on the same avenue. On the other hand, two dealerships left the Manoel Goulart extension in the direction of Avenida Joaquim Constantino.

Map 7

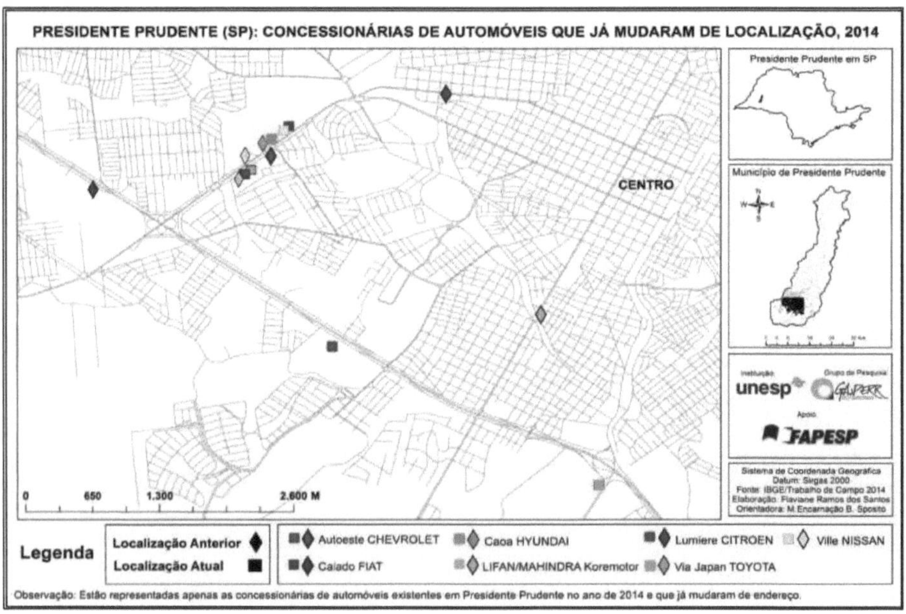

Source: IBGE (2010) and interviews.

With regard to the property where the dealerships are located, we found that nine of them operate in rented buildings, while only Autoeste, Aurora, Liane and V.M. *Center,* i.e. four, own their respective buildings, according to map 8.

Map

PRESIDENTE PRUDENTE (SP): ANO DE INSTALAÇÃO DAS ATUAIS CONCESSIONÁRIAS DE AUTOMÓVEIS

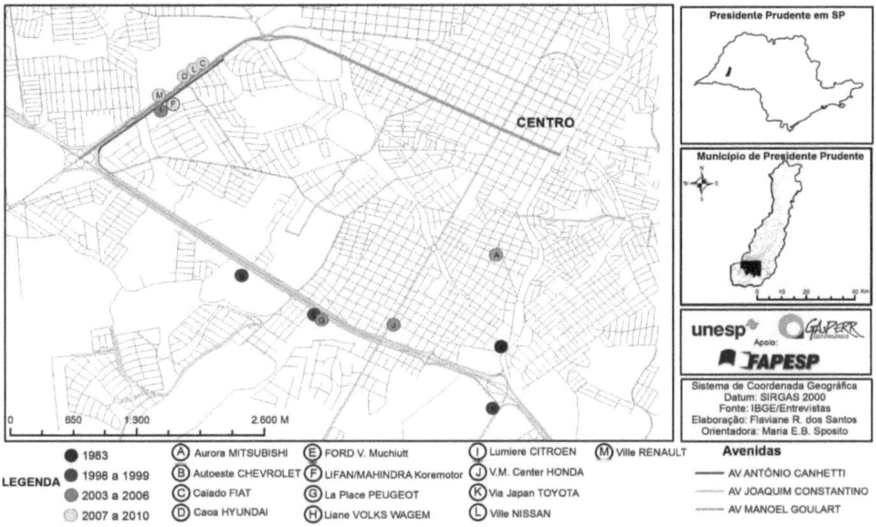

Source: Interviews.

Another important issue is that, with the exception of Autoeste, which is located in its own building on Avenida Joaquim Constantino, the other three dealerships that operate in their own buildings are scattered around the city, on Avenida Brasil, Coronel José Soares Marcondes and Rua Antônio Rodrigues, In this case, the choice to set up and, above all, remain in their own premises was decisive, indicating that, among the agents of urban space production, the role of the landowner still has its weight in locational decisions. However, in the case of Aurora and V.M. Center, the intention is to move in the near future, given the impossibility of growing in their current locations.

Well, we've already done three renovations, we've rented buildings around us to increase the space and we still need more space. So our plans for the future are expansion, for sure. Relocation and expansion. [Interviewee: Where are you thinking of moving from here? Any ideas?] To the highway itself, to the side of the highway (Sales Manager at V.M. Center Honda).

In fact, the location was chosen precisely because the land was *already* there, there was already a building here, so, in reality, we just took advantage of the structure. We didn't do a market analysis to find out what would be ideal, the ideal would be on the highway, for example, where Autoeste is today, near the city exit... or at SP-270, which is Raposo Tavares, that would bring more visibility than here, but the investment there with the structure we already had here, at the tip of the pencil, made it worthwhile to stay here, at least for now. But we plan to expand, and the idea is that one day we'll have space to build a dealership closer to the highway... Then we'll move to another location, because here we can no longer afford to expand (Sales Manager at Aurora Mitsubishi).

Considering that the average number of cars sold (Graph 9) is linked to the size of the dealership, as well as its notoriety on a national scale and its acceptance by city dwellers on a local and regional scale, we found well-balanced sales averages among the city's dealerships. In other words, five of them said that their average sales did not reach 50 cars a month. On the other hand, of the remaining eight, four have averages of between 50 and 100 cars/month and the other four sell more than 100 cars a month.

Graph 9

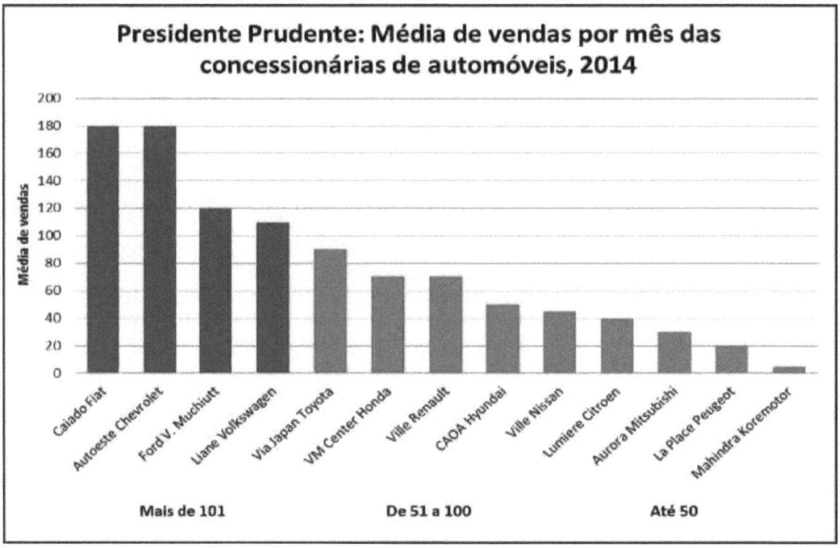

Source: Interviews.

However, as the interviewees themselves pointed out, these averages vary according to the time of year, with the highest averages occurring in the months around the vacations, i.e. June/July and November/December, while the worst averages are exactly in the months after the vacations, i.e. August/September and January/February. The interviewees' explanation for this is that people who are going to buy a car usually choose to buy them before going on vacation or use the money they have been saving throughout the year to make the purchase. In addition, the drop in sales at the beginning of the year is generally due to the increase in expenses, such as taxes related to the home or vehicles, such as IPTU (Urban Property Tax) and IPVA (Motor Vehicle Ownership Tax), respectively, or even debts related to children's school, such as materials, uniforms and, above all, registration fees, among other expenses.

Sales are generally higher from August to November. Generally, people want to change their car for a trip at the end of the year and already want to go with a new car. So it's the most favorable time of year, the time that sells the most. [The beginning of the year now. IPVA, children's school, IPTU, so everyone takes out insurance at the beginning of the year. (Salesman at Autoeste Chevrolet).

Normally this period of November and December is better. The shortest would be between the end of December and the beginning of January, because the car manufacturers stop, the distribution of cars from there to here for delivery takes a little longer, so the market is a little slower during this period and it's also when people go on vacation. Another time is July, because it's also vacation time, and people get ahead of themselves, they buy their cars to take their vacations. (Renault saleswoman).

As far as employees are concerned, we found that dealerships are organized into various departments, such as new and used car sales, technical assistance, parts and accessories, financing and insurance, among others.

In the **new and used** car **department**, just like salespeople in general, car salespeople are paid in the form of a commission, in addition to a flat rate close to the minimum wage. This form of remuneration exerts great pressure on the sales teams, because if the sale is not made, the salesperson does not receive his commission, and his salary becomes only the fixed amount (LIMA, 2009).

In addition, new car buyers usually want to sell their used cars, by offering them as part payment. Used cars accepted as part payment when selling new cars go into the dealership's stock and are sold on to another type of consumer. This way,

The used or semi-new vehicle department at dealerships works with a team of appraisers, who are responsible for establishing the purchase price of the used vehicle when it is part of the payment for the new vehicle being purchased. It is at this point that a conflict of interest arises between the sellers and the owners, since the customer wants the highest price for the sale of their used vehicle, while the dealership wants to pay the lowest price (LIMA, 2009, p.92-93).

In the department responsible for after-sales service or vehicle maintenance, the employees must be at the customer's disposal when they want to make any kind of repair to their car (LIMA, 2009). It is usually made up of a team of technical consultants, a workshop manager, mechanics and mechanics' assistants.

Unlike most mechanical workshops (not authorized by car manufacturers), which specialize in only one type of service, a car dealership provides services in all areas of technical assistance. These are Mechanical (carries out mechanical repairs); Electrical (carries out repairs to electrical and electronic components); Bodywork (repairs damage and replaces parts caused by collisions); Paintwork (paints the parts repaired and replaced in the bodywork); Bodywork (repairs and replaces the internal parts of vehicles, such as glass, rubber, upholstery, etc.) (LIMA, 2009, p.96).

According to the same author, and as found in the interviews, the majority of service customers only go to the dealership during the warranty period, because once the warranty expires, consumers migrate to independent workshops in search of better prices.

The **workshop and parts department** sells spare parts for vehicles and is strategically linked to the technical assistance department.

The Parts Department requires more sophisticated management, as it involves several thousand variable items, such as forecasting future demand, safety stock, assortment, obsolete level management and purchasing cycles (LIMA, 2009, p.98).

Another characteristic of this department is that the part is a consumer product without any impulse appeal. In other words, "you don't buy a part because a good promotional campaign has generated a 'need' in the consumer's mind; the demand for the part arises because the vehicle needs to be repaired." (LIMA, 2009, p.98). In addition, the parts department usually has a team that sells parts to other workshops, so that they can serve the brand's customers who have chosen to have their vehicle repaired outside the authorized dealership.

And finally, the **financing and insurance department**, which has a very specific function: to encourage car buyers to negotiate their financing and insurance with the dealerships' "partner" finance and insurance companies.

Thus, taking into account all the departments mentioned above, we see that the average number of employees at most dealerships is up to 50, while three have between 50 and 100 and only two have more than 100, as shown in graph 9. Another important issue is that, given the greater demand, it is the after-sales (service) sector that concentrates the largest number of employees at car dealerships.

Graph 10

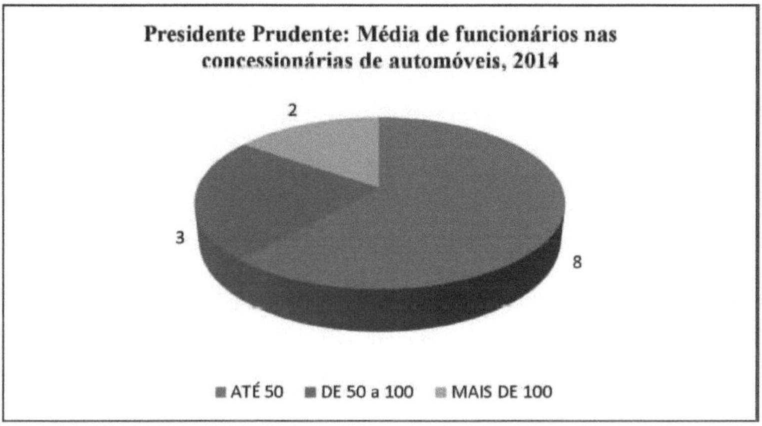

Source: Interviews.

It's important to note that the number of employees (Table 3) is related to the average number of cars sold and, consequently, the size of the dealership. Thus, the three dealerships with between 50 and 100 employees (Caiado *Fiat, Ford* V.Muchiutt and *Via Japan Toyota*), as well as the two with more than 100 (Liane *Volkswagen* and Autoeste *Chevrolet*) are also the ones with the most widely accepted vehicles on the car market.

Table 3 - Average number of employees at car dealerships, 2014

CONCESSION	**AVERAGE NUMBER OF EMPLOYEES**
Autoeste Chevrolet	115
Liane Volkswagen	100
Caiado Fiat	87
Ford V. Muchiutt	80
Via Japan Toyota	55
Lumière Citroen	40

V.M. Center Honda	36
CAOA Hyundai	35
Ville Renault	32
Ville Nissan	30
Aurora Mitsubishi	28
La Place Peugeot	14
Mahindra Koremotor*	5*

* The interview was conducted when Koremotor was in its old location and did not have after-sales services, so the number of employees refers only to the three salesmen, the administrative manager and the car washer. Today, in its new building, this number is higher.

Org. Flaviane Ramos dos Santos. Source: Interviews.

Regarding the average frequency of the arrival of shame trucks at car dealerships (Graph 10), we can see that it is also related to the average number of vehicles sold in the month. In other words, those dealerships that sell the most vehicles during the month receive more vehicles to replenish their stock, such as Caiado *Fiat, Ford* V.*Muchiutt*, Liane *Volkswagen* and Via *Japan Toyota*, which receive new vehicles around three to five times a week. The opposite is also true, i.e. those dealerships that sell fewer cars during the month receive the shame trucks less often.

Graph 11

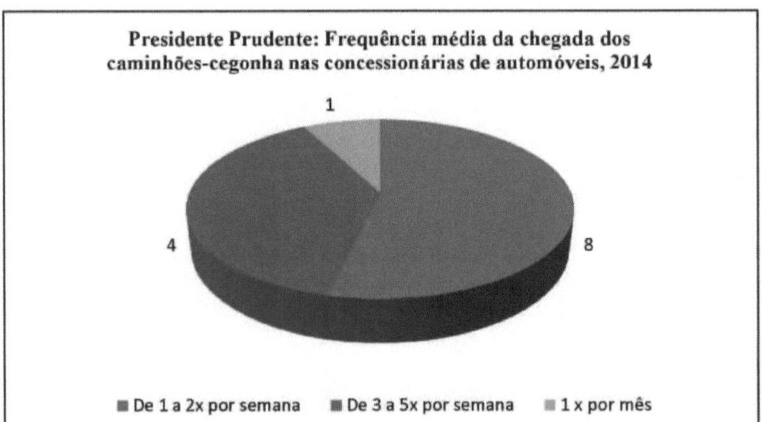

Source: Interviews.

With regard to the interviewees' opinions on proximity to other dealerships (Graph 11), the majority of them said that this was a positive aspect, for two main reasons: a) it speeds up the consumer's return to the dealership, because it takes them less time to research prices and; b) it makes it easier for the consumer to research prices, which is a very important factor in situations of occasional consumption, such as buying a car, as in the following interview excerpts.

Graph 12

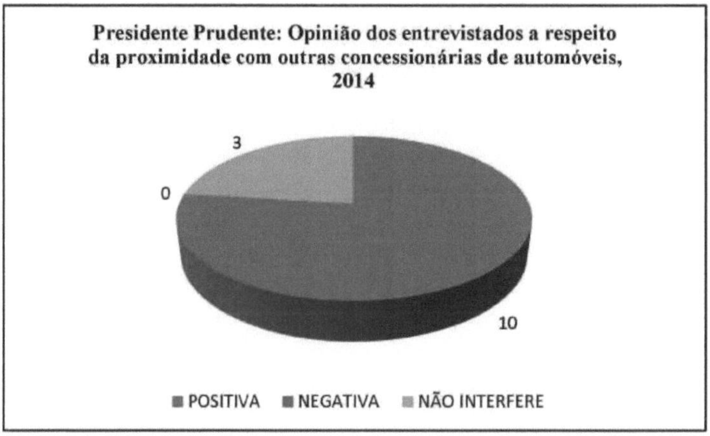

Source: Interviews.

For us here at Renault, in fact, because nowadays people have a lot of information. I think that being close speeds up the process of the customer returning. Let's say, if they go somewhere further away, it will take them longer to come back here if they really like my proposal, if they like the service or the price and everything else. So I think it makes it easier, because it's quicker for her to come back, because nobody comes straight in to buy a car. I arrived, I looked, I liked it, I bought it. She'll do some price research on the competition. So this, in my opinion, will speed up the customer's return if they really like the price, the service and the product (Saleswoman at Renault).

Positive. As long as you have competitors nearby, you're bound to end up with a larger flow of customers. This can be good, but it can also be bad, depending on the ability of the dealership, its management and everything else, to be able to absorb this number of customers and really bring them to your brand, in other words, to do business. Anyone with bad service is already out of the market (Sales Manager at La Place Peugeot).

It's good to be close, because today in the market everyone does research. So the closer you are, the easier it is for someone to be in a place where they *already* know there's one next door, they go to one place, they automatically go to the other, and so much so that the trend now is for all the dealerships to come to the side of the highway. So much so that we have other dealerships on the avenue itself here, we have other dealerships that are close by, so the ease of being close by is the advantage of the search, so it's better for the consumer who can search in several dealerships in less... in a short time (Autoeste Chevrolet salesperson).

Meanwhile, the three interviewees who said that proximity to competitors had no effect on their profitability were precisely the representatives of *Ford, Volkswagen* and *Fiat*, i.e. V. Muchiutti, Liane and Caiado, respectively. This means that whether or not you are close to another dealership is not a preponderant factor, given that these are the three most consolidated brands on the market, both locally and regionally and nationally, as shown in capital 1.

FINAL CONSIDERATIONS

According to the results obtained from the research carried out between August 2013 and January 2015, which served as a reference for this book, it was found that it is in the south-southwest sector of the city of Presidente Prudente that there is a concentration of dealerships and other companies associated with commerce and services focused on automotive transportation, and that their locational choices are, This serves as an example of the process of restructuring the city and the valorization of the south-southwest quadrant, resulting from the occupation of "urban voids" created in the 1970s and 1980s, accompanied by the new "wave" of expansion of the urban fabric.

Studying the spatial logics of companies that sell automobiles offers elements for understanding the links between urban and city restructuring. On an inter-urban scale, it is important to note that these are companies that represent the large international automakers operating in the country and that develop commercial and spatial logics according to the strategies and norms established by these transnationals. On a city scale, there have also been major changes, because while there used to be a tendency for dealerships to concentrate on the commercial and service axes that unfolded from the main center, more recently they have been established on more distant axes.

As a result of this research, it was possible to come up with some explanations to understand the strategies and interest of car dealerships in setting up in the vicinity or directly on the extension of Avenida Manoel Goulart, Avenida Antônio Canhetti and Avenida Joaquim Constantino, important axes in terms of the flow of people and vehicles in the city of Presidente Prudente. Thus, some of the main strategies identified through interviews and fieldwork were:

- The need to move away from the main center in order to avoid traffic jams and parking difficulties, from which we can see which clientele is primarily targeted and served, while at the same time looking for relatively peripheral locations that offer areas with larger and cheaper plots of land.
- The importance of locating on roads with a large flow of vehicles and people who

pass through the avenues on a daily basis. In the case of the extension of Avenida Manoel Goulart, these are important intra-urban flows towards the city's most populous housing estates, such as Ana Jacinta, and in the case of Avenida Joaquim Constantino, they are also inter-urban flows;

- Proximity to competing car dealerships, which allows consumers to return to the dealership more quickly, because it takes them less time to go to competitors and do their price research, as well as making it easier for them to do their research, which is a very important moment since buying a car is an occasional consumption situation;

- The relative proximity to residential areas of high and medium occupational standards generates symbolic benefits due to the value that these closed residential areas give to their surroundings, as well as revealing the concessionaires' desire to reach customers with greater purchasing power, whether they live in closed residential areas or not, and gives these avenues levels of socio-economic specialization.

> In many cases, and in others, this specialization translates into a search for the higher purchasing power segments of the market, aimed at a "special" clientele, which in this sense becomes distinguished and therefore spatially separated from other clientele. [...] insofar as they are aimed at customers with greater purchasing power, they offer access facilities for individual transportation, security [...] in short, "social distinction" (SPOSITO, 1991, p. 11 and 12).

- Proximity to the city's entry and exit routes is also relevant, not only from the point of view of the consumer market, in that it facilitates access for people from other cities, but also from the point of view of the supply of goods to these dealerships, after all, it is not by any routes that car transporters, popularly known as "stork trucks", can easily move around or be parked;

In addition, as can be seen in Appendix IV, these locational strategies adopted by car dealerships in Presidente Prudente are not unique to this city. On the contrary, to a greater or lesser extent, these strategies occur in various cities, including cities in other countries, as we were able to verify in the city of Porto in Portugal during the research internship abroad, funded by FAPESP .[24]

---

24 Process: 2013/23185-6.

REFERENCES

Yearbook of the Brazilian Automotive Industry 2013. Sao Paulo: ANFAVEA, 2014. Available at:< http://www.anfavea.com.br/anuario.html> Accessed on: October 2014.

Yearbook 2013: The performance of automotive distribution. Sao Paulo: FENABRAVE, 2014. Available at:

<http://www3.fenabrave.org.br:8082/plus/modulos/listas/index.php?tac=indices-e-numeros&idtipo=6&layout=indices-e-numeros> Accessed: October 2014.

BARATA-SALGUEIRA, Teresa. From the center to multiple centralities. In: FERNANDES, José Alberto Rio and SPOSITO, Maria Encarnaçao Beltrao. **The new life of the old center:** in Portuguese and Brazilian cities. P. 13-29, 2013.

BRAZIL. National Traffic Department (DENATRAN). Vehicle fleet by municipality 2002, 2005, 2006, 2007, 2008, 2009, 2010, 2011, 2012, 2013, 2014.

BRAZIL. Brazilian Institute of Geography and Statistics (IBGE). Demographic Censuses 1970, 1980, 1991, 2000 and 2010.

BUFFON, Marciano. The IPI Reduction in the Automotive Sector: Who Wins? Revista Tecnicouro, Novo Hamburgo - RS, ed. 260, p. 32, ano 32/n° 3, Apr 2011. Available at: <https://justicafiscal.wordpress.com/category/politica-fiscal/> Accessed in: November 2013.

Letter from the Association of Automotive Vehicle Manufacturers, December 2014. Sao Paulo: ANFAVEA, 2014. Available at: < http://www.anfavea.com.br/cartas/carta343.pdf> Accessed: December 2014.

CAMAGNI, Roberto. **Urban Economy.** Barcelona: Antonio Bosch, 2005.

CATELAN, Màrcio José. Urban heterarchy and inter-scalar spatial interactions: An analytical proposal for urban network studies. In: National Symposium on Urban Geography, 13°, 2013, Rio de Janeiro. Available at: <http://www.simpurb2013.com.br/wp-content/uploads/2013/11/marcio.pdf> Accessed on: November 2014.

CENTER FOR STUDIES AND MAPPING OF SOCIAL EXCLUSION FOR POLICIES

PÙBLICAS - CEMESPP (Presidente Prudente, SP). **Atlas of social exclusion in the interior of São Paulo**. Presidente Prudente, 2003. 1 CD ROM.

CORREA, Roberto Lobato. **The urban space**. Ed. Atica, Sao Paulo, 1989.

. Reflections on the recent dynamics of the Brazilian urban network. In: CORRÊA, Roberto Lobato (Org.) **Estudos recentes sobre a rede urbana** Ed. Bertrand Brasil. Rio de Janeiro. 2006.

. On social agents, scale and the production of space: A text for discussion. In: CARLOS, Ana Fani Alessandri; SOUZA, Marcelo Lopes; SPOSITO, Maria Encarnaçao Beltrao. (Org.) **The production of urban space**: Agents and processes, scales and challenges. Sao Paulo: Contexto, 2011.

COSTA, Rodrigo Morem da. The automobile industry in Brazil and the crisis in Argentina. Available at:< http://carta.fee.tche.br/article/a-industria-automobilistica-no-brasil-e-a-crise-argentina/> Accessed: December 2014.

DAL POZZO, Clayton Ferreira. Socio-spatial fragmentation: Analysis of the socio-spatial practices of self-segregated subjects in Presidente Prudente/SP. Presidente Prudente, 2008. Final paper (BA in Geography). - São Paulo State University, Faculty of Science and Technology.

FERNANDES, José Alberto Rio. Many lives have the center and several centers have the life of a city. In: FERNANDES, José Alberto Rio & SPOSITO, Maria Encarnaçao Beltrao (Org.) **A nova vida dos velhos centros** 2013. P.31-44.

GRANDE, M. M. Is the distribution of new cars changing? A study based on a *survey* and qualitative research in dealerships. 2004. Thesis (Doctorate in Production Engineering). Faculty of Engineering, Polytechnic School, University of Sao Paulo - USP. Sao Paulo, 2004.

JOHNSON, James. **Urban Geography**. Barcelona: Oikos-tau, 1974.

KON, Anita. Industrial localization, polarization and regionalization. In: **The industrial economy**. Sao Paulo: Nobel, P.157-178, 1994.

LIMA, Marcos Antônio de. Contribution to understanding the operating model of car dealerships in Brazil and the use of F&i - *finance and insurance* : A case study in a large FIAT dealership network. Sao Caetano do Sul/SP, 2009. Dissertation (Master's in Administration). Municipal University of Sao Caetano do Sul - USCS.

TELESP 175 Telephone Directory for Presidente Prudente and Region. EPIL Publishing House, 1987/1988.

TELESP 176 Telephone Directory for Presidente Prudente and Region. EPIL Publishing House, 1998/1999.

Telephone directory of Presidente Prudente and Region. EPIL Publishing House, 2008/2009.

Telephone directory of Presidente Prudente and Region. EPIL Publishing House, 2013/2014.

LOPES, Adcmil Lucio. The deconstruction of the Brazilian automobile industry formed during the JK government by the sectoral policy of the FHC government. UNIARA Magazine n°17/18 2005/2006. Available <

http://www.uniara.com.br/revistauniara/pdf/17/rev17completa_15.pdf> Accessed: November 2013.

LUNA, Cacilda. Incentives for the automotive industry: an old controversy. Jornal Gente da FEA (Faculty of Economics and Administration) of the University of Sao Paulo (USP), 2013. Available at:< http://www.fea.usp.br/noticias.php?i=1095> Accessed: November 2013.

OLIVEIRA, Célio Roberto Isidio de. A study on the measurement of organizational performance in car dealerships located in the metropolitan region of Recife. Recife/PE, 2006. Dissertation (Master's in Accounting) Multinstitutional and interregional postgraduate program at University of Brasilia (UnB), Federal University of Paraiba

(UFPA), Federal University of Pernambuco (UFPE) and Federal University of Rio Grande do Norte (UFRN).

PEREIRA, Silvia Regina. Expansion and internal structuring of the urban space of Presidente Prudente/SP. In: Subcentros e condições de vida no Jardim Bongiovani e Conjunto Habitacional Ana Jacinta em Presidente Prudente/SP. Presidente Prudente, 2002. Dissertation (Master's Degree in Geography). São Paulo State University (UNESP).

PINTAUDI, Silvana Maria; FRÙGOLI JR, Heitor. **Shopping Centers:** Space, culture and modernity in Brazilian cities. Sao Paulo: Editora UNESP, 1992.

PRESIDENTE PRUDENTE. History of commerce and industry in the city. <http://camarapprudente.sp.gov.br/historia/hist_oeste/cidades/pprudente/comercio.html> Accessed on: May 2015.

PRESIDENTE PRUDENTE. City Hall. General plan: Zoning of land use and occupation. occupation. Available at: <http://www.presidenteprudente.sp.gov.br/site/central_mapas.xhtml;jsessionid=5A0D9394475682AE806BF7775D2E9F22> Accessed: January 2015.

SANTOS, Milton. **Manual of Urban Geography.** Sao Paulo: Editora Hucitec, 1989.

SCHNEIDER, Sérgio; SCHMITT, Claùdia Job. The uses of the comparative method in the social sciences. Cadernos de Sociologia, Porto Alegre, v. 9, p. 49-87, 1998. Available at:<http://ucbweb2.castelobranco.br/webcaf/arquivos/23762/2196/Texto_auxiliar_para_consultar_O_uso_do_metodo_comparativo_nas_ciencias_social.pdf> Accessed on: March 2014.

SAVÉRIO SPOSITO, Eliseu. Cities. In: **Networks and cities**. Sao Paulo: Editora UNESP, 2008.

SOBARZO, Oscar Alfredo. **The spaces of segmented sociability:** The production of public space in Presidente Prudente. Presidente Prudente: 2004. Thesis (Doctorate in Geography) - Universidade Estadual Paulista, Faculdade de Ciências e Tecnologia.

SOUZA, Claudio Luis de; et al. ABC strike causes Volkswagen to lose 1,400 cars per day. Available at:< http://carros.uol.com.br/noticias/redacao/2015/01/09/greve- no-abc-faz-volkswagen-perderder-1400-cars-por-dia.htm> Accessed on: January 2015.

SOUZA, Marcelo Lopes de. The city from the inside. In: **ABC do desenvolvimento urbano**. Rio de Janeiro: Bertrand Brasil, 2003. P.63-80.

SPOSITO, Maria Encarnaçao Beltrao. **The ground in Presidente Prudente: The logic of urban territorial expansion.** Rio Claro. 1983. Dissertation (Master's Degree in Geography). 230 p. - Faculty of Science and Technology Paulista State University, Rio Claro.

. The center and the forms of expansion of urban centrality. In: **Revista Geografia**. Ed. UNESP, Sao Paulo, 1991, p. 1-17.

. New commercial forms and redefinitions of intra-urban centrality. In: SPOSITO, Maria Encarnaçao Beltrao. (Org.) **Textos e contextos para a leitura geogràfica de uma cidade média**. Presidente Prudente: UNESP, 2001a.

. Medium-sized cities and contemporary economic contexts. In: SPOSITO, Maria Encarnaçao Beltrao. (Org.). **Urbanization and cities:** Geographical perspectives. Presidente Prudente. 2001b.

. The production of urban space: Scales, differences and socio-spatial inequalities. In: CARLOS, Ana Fani Alessandri; SOUZA, Marcelo Lopes; SPOSITO, Maria Encarnaçao Beltrao. (Org.) **The production of urban space**: Agents and processes, scales and challenges. Sao Paulo: Contexto, 2011.

. Economic logics and contemporary spatial practices: medium-sized cities and consumption. Presidente Prudente: thematic project approved by FAPESP, 2012.

. Centers and centralities in Brazil. In: FERNANDES, José Alberto Rio & SPOSITO, Maria Encarnaçao Beltrao (Org.). A **nova vida dos velhos centros** 2013. P.45-59.

SPOSITO, Maria Encarnaçao Beltrao; GÓES, Eda Maria. **Closed spaces and cities**: Urban insecurity and socio-spatial fragmentation. Sao Paulo: UNESP publishing house, 2013.

VILLAÇA, Flàvio. **Intra-urban space in Brazil.** Sao Paulo: Studio Nobel, 2001.

VOLKSWAGEN. Volkswagen factories in Brazil. Available at: <http://www.vw.com.br/pt/institucional/VolkswagenBrasil/fabricas.html> Accessed on: January 2015.

ANNEXES

ANNEX I - Data taken from the "Anuàrio da indùstria automobilistica no Brasil" (2013 edition) and used in the graphs in chapter 1.

YEAR	NET SALES (billions of U$S)	PARTICIPATION In GDP INDUSTRIAL	TOTAL PRODUCTION	TOTAL EXPORTS	EMPLOYMENT
1957	-	-	1.166	-	9.773
1958	-	-	3.831	-	19.248
1959	-	-	14.495	-	29.323
1960	-	-	42.619	-	38.410
1961	-	-	60.205	-	37.404
1962	-	-	83.875	-	48.523
1963	-	-	94.764	-	43.994
1964	-	-	104.710	-	44.414
1965	-	-	113.772	-	49.456
1966	-	-	128.821		50.662
1967	10.924	12,6%	139.260	-	46.396
1968	11.074	13,8%	165.045	-	60.437
1969	13.594	14,6%	244.379	3	61.059
1970	16.164	14,6%	306.915	52	65.902
1971	17.892	15,5%	399.863	656	71.406
1972	24.430	16,2%	471.055	6.611	80.430
1973	28.534	17,3%	564.002	13.891	96.099
1974	32.880	18,6%	691.310	47.591	104.072
1975	36.868	20,6%	712.526	52.629	104.556
1976	36.580	18,1%	765.291	62.079	112.429
1977	36.454	17,1%	732.360	56.636	111.514
1978	40.050	16,8%	871.170	77.388	123.974
1979	36.893	15,0%	912.018	76.486	127.081
1980	34.829	13,3%	933.152	115.482	133.683
1981	26.688	11,5%	585.834	157.228	103.992
1982	30.601	12,8%	672.589	120.305	107.137
1983	28.969	12,6%	748.371	132.804	101.087
1984	28.513	12,4%	679.386	151.962	107.447
1985	36.220	14,3%	759.141	160.626	122.217
1986	34.724	12,3%	815.152	138.241	129.232
1987	37.324	13,5%	683.380	279.530	113.474
1988	42.137	15,0%	782.411	226.360	112.985
1989	31.102	10,9%	730.992	164.885	118.369
1990	28.639	10,6%	663.084	120.377	117.396
1991	30.042	10,8%	705.303	127.153	109.428
1992	37.191	13,8%	815.959	243.126	105.664

1993	43.079	14,6%	1.100.278	249.607	106.738
1994	45.723	14,8%	1.248.773	274.815	107.134
1995	43.842	13,1%	1.297.467	189.721	104.614
1996	47.361	13,9%	1.458.576	211.565	101.857
1997	51.930	14,7%	1.677.858	305.647	104.941
1998	49.807	14,8%	1.254.016	291.788	83.049
1999	41.383	12,6%	1.109.509	204.024	85.100
2000	45.063	13,0%	1.361.721	283.449	89.134
2001	45.818	13,5%	1.501.586	321.490	84.834
2002	43.402	12,9%	1.520.285	369.925	81.737
2003	42.039	12,5%	1.505.139	440.957	79.047
2004	52.009	14,6%	1.862.780	603.052	88.783
2005	56.133	14,6%	2.011.817	684.260	94.206
2006	60.110	15,1%	2.092.003	635.851	93.243
2007	71.715	17,3%	2.391.351	588.341	104.274
2008	76.245	17,8%	2.545.729	558.207	109.848
2009	77.259	18,7%	2.575.418	373.747	109.043
2010	86.066	19,2%	2.825.974	616.125	119.392
2011	84.980	18,9%	2.505.241	405.575	125.972
2012	83.676	18,7%	2.591.348	308.693	129.997
2013	-	-	2.723.411	397.554	131.595

Source: ANFAVEA (2013).

Org. Flaviane Ramos dos Santos.

ANNEX II - Data on the car fleet in Brazil and in Presidente Prudente taken from the DENATRAN website and used in chapter 1.

Total car fleet		
	BRAZIL	**PRESIDENT PRUDENTE**
2002	22.486.611	46.245
2004	23.757.230	47.364
2006	26.416.664	51.860
2008	30.021.842	57.477
2010	34.696.707	65.431
2012	40.029.320	75.047
2014	45.672.929	83.486

Source: DENATRAN.

Org. Flaviane Ramos dos Santos.

MIX
Papier aus verantwortungsvollen Quellen
Paper from responsible sources
FSC® C105338

Printed by Books on Demand GmbH, Norderstedt / Germany